Human Factors in Industrial Design

The Designer's Companion

John H. Burgess

TAB Professional and Reference Books

Division of TAB BOOKS Inc.

Blue Ridge Summit, PA

A Petrocelli Book
FIRST EDITION
FIRST PRINTING

Library of Congress Cataloging in Publication Data

Burgess, John H.
    Human factors in industrial design : the designer's companion / by
John H. Burgess.
        p. cm.
    Bibliography: p.
    Includes index.
    ISBN 0-8306-3356-1
    1. Design, industrial. 2. Human engineering. I. Title.
TS171.4.B87 1989
745.2—dc20   89-33594
                                                                      CIP

TAB BOOKS Inc. offers software for sale. For information and a catalog, please contact
TAB Software Department, Blue Ridge Summit, PA 17294-0850.

Questions regarding the content of this book
should be addressed to:

    Reader Inquiry Branch
    TAB BOOKS Inc.
    Blue Ridge Summit, PA 17294-0214

Contents

CONTENTS

III

———————— Human Factors Design Data ————————

IV

———————— Product Improvement ————————

V

———————— New Products ————————

VI

———————— Total   Systems ————————

VII

———————— Design Evaluation ————————

Contents

VIII

———————————— Product Safety ————————————

———————————— Appendices ————————————

Preface

Human factors material tailored to a designer's orientation has seemed to me for some time to be a most needed contribution in the human factors literature. Human factors specialists usually have a background in experimental psychology or industrial engineering. Both of these fields rely heavily on statistics and mathematical analytical techniques. Psychologists are also quite extensively trained in human anatomy, physiology and neurology. Such background components in either—or both—mathematics and biological science are generally lacking in a designer's curriculum.

While a human factors specialist emphasizes scientific validity and reliability of data in human performance, the designer generally has a different emphasis. Industrial designers may concentrate more on the aesthetic characteristics and marketability of a product, whereas the engineer's emphasis is more on the functional performance of the equipment. When more specific emphasis is to be placed on human factors, as is frequently the case in military projects, human factors experts are called in. In other than military projects, it has been my experience that the human factors aspects of a project are simply glossed over. Human factors design decisions are then made on an intuitive, or, if you will, on a quasi-random basis. The latter, indeed, may be the case in perhaps 80 to 90 percent of all nonmilitary design projects.

In a typical engineering graphics textbook, as a case in point, there was virtually no indexing nor referencing for human engineering. The book stated simply that it is the industrial engineer who is responsible for a human factors component: ". . . Not only must the industrial engineer have a sound knowledge of the principles of engineering, but he must possess an understanding of the human factor . . ." (*Engineering Graphics* by R. Hammond, et al., p.6).

In teaching human factors to third-year design students, I found both faculty and students to be keenly interested in the human factors aspects of design. However, a generic human factors background was lacking in their training. Consequently, they tended to find the human factors material in typical human factors texts difficult to assimilate. The reason for this seems generally to be in the differences in emphasis; while human factors texts are written for specialists, a designer requires simplified and expedient techniques for immediate design applications.

The material I have devised is thus oriented primarily to a design emphasis. The human user of products is described in terms that a designer might easily understand—as a kind of black box with definitive perceptual sensitivities and processing characteristics and output modes of response.

The text is best suited as a companion volume required as an integral part of substantive design courses. While a separate human factors course for designers, emphasizing the validity of data and research applications, will enhance the depth of understanding, an ongoing curriculum requirement for human-factors applications may be the most pertinent. Indeed, simply teaching a human factors course to designers, without an essential design emphasis, may tend to isolate the subject matter. The student may get the unfortunate message that such considerations are not really all that relevant to design. The integration of human factors into design courses should no longer be a compartmentalized study nor made to seem to be simply an unnecessary embellishment in design. A human factors component in virtually any engineering or industrial design project is paramount to good design and consumer satisfaction. A real need emerges to translate the human factors imperative into operational design terms and such a premise has been made the central theme of this book.

The material has been sectionalized in the text to accommodate design curricula that proceed from simple to complex design projects. Thus, human factors analytical techniques are presented, in operational terms and by way of examples, that can be applied to a gradation of complexity—from simple hand tools to complicated systems. Basic human engineering data and cases of application are included, as well as references for acquiring more unique design data when needed.

In the process of developing an effective pedagogical procedure, I found that engineering and industrial design students were particularly receptive to carrying out design exercises. This particularly became the case once they were familiarized with human factors analytical methods and principles of application. Thus, each section calls for detailed practical human factors analyses of activities and tasks. Human engineering data applications are called for in product improvement and initial product design as well as for the more complex systems analyses.

J. H. Burgess
Syracuse, N.Y.

Introduction

Human factors is a general term that applies when any consideration is given in design for the users of a product or piece of equipment. It also applies when a Human Operator (HO)[1] is used to fulfill operating requirements in more complex systems. The term also refers to what is called "human engineering," "engineering psychology," or "ergonomics."[2]

The term was first used in the 1940s during World War II, though, of course, generic types of human factors concerns have operated for centuries. In recent times, Charles Babbage wrote a book entitled *Economy of Machinery and Manufacture* (1832) in which he laid out methods for making jobs easier and more economical. Adam Smith also did this a half-century earlier in his book, *Wealth of Nations* (1776). Later, the industrial engineer Frederick Taylor, at the Midvale Steel Company in Philadelphia, studied the human factors of hand tools, the object being to increase human productivity. The famous Gilbreths studied motion and economy factors in work during the 1920s and 30s, and their methods became extensively used for improving industrial effici-

---

1. The term "HO" will be used extensively throughout the text, generically referring to both male and female human users. Functional characteristics of the HO, with pertinent capabilities and limitations, are described in the chapters that follow.

2. deMontmollin and Bainbridge see a difference between the American definition of ergonomics and that of the British and Europeans. In America, they see it as referring to biomechanics and the work environment. In Europe, they claim, "ergonomics" is synonymous with what Americans mean by "human factors." (Refer to the *Human Factors Society Bulletin*, June, 1985, p. 1ff.)

ency. The psychologist Hugo Munsterberg first used psychological testing in industry for personnel selection.

Human factors studies became extensive during World War II when poor human engineering had cost both sides many lives and resulted in ineffective weapons. Such names as Helson, Craig, Ellson, Fitts, Flexman, Grether, Kaufman, Loucks and Mitchell were among the many British and American engineering psychologists who improved weapons efficiency during the war.

After the war, engineering psychologists participated in systematizing machine systems for more effective human operator use of weapons in the "cold war." Military operations proved to be most successful when human operator capabilities were considered throughout the entire system development program. Early work at the Navy, Army Signal Corps and Air Force human engineering laboratories provided voluminous data. Alphonse Chapanis at John Hopkins University wrote among the first in a series of textbooks summarizing and updating the human engineering state of the art.

Human engineering is commonly recognized to be an important part of the development process in most aerospace and military products. Human factors in the design of civilian products and industrial operations is only beginning to receive such attention, largely due to publicized incidents where human error has been found to be responsible for accidents. For example, the Three Mile Island nuclear plant incident in Pennsylvania proved to be a most costly and dangerous episode that was traceable to poor human engineering.

In modern industry, accidents have been found to be attributable to the poor human engineering of equipment. The neglected human factor has resulted in hazardous operations, inconveniences and inefficiencies in both product and industrial operations. With the advent of increasing numbers of female workers in the workplace, the female human factor has also been sorely neglected; women are subjected to ill-fitting face masks, coveralls and footwear, as well as tools and equipment only poorly designed for them. All this, of course, means discomfort and poor morale for the worker, and poor productivity for the industry.

Product life is also often severely compromised, partly due to the lack of good ergonomic or human factors design. The early digital watches, for example, were so difficult to set and reset that many consumers simply gave up in disgust.

Numerous aircraft accidents continue to be traceable to human error and, in turn, the lack of good human engineering. Automobile mishaps are attributable to any of a number of poor human engineering design features including the roadway subsystem, traffic regulatory and maintenance subsystems, and the vehicle subsystem itself.

Nuclear power plants, chemical factories, coal mining machines, and a myriad of product-safety deficiencies can be found to be associated with, or directly related to, the lack of human factors considerations, or just poorly human engineered user interfaces.

Human operator interactions occur in almost all steps in the life cycle of a machine or piece of equipment, and industrial design and engineering must assume direct responsibility for how these interactions occur. The machine may simply be an appliance or other domestic product, or it can be a complex operational system, but the responsibility is still there to provide the most efficient human engineering design features possible. These human interfaces should be considered in every design project. All aspects of interaction with the machine should be determined for a valid specification of display and control requirements. The condition of the ambient and machine-generated environments should also be determined. From such data and from special design studies, sound human engineering applications can be made.

Engineering and industrial design students must be sensitized to all aspects of human interfacing in their product designs. This will assure that a well-rounded and mature perspective on human factors will be incorporated. By assuming such responsibility at the beginning of design, costly human error and poor human performance or productivity can later be prevented.

Human Factors in Industrial Design (ID)

Industrial designers are taught the basics of human factors methods and data applications. However, in some ID curricula, human factors instruction is limited to a kind of general admonition, viz., to be sure to consider the "human" aspects of the product. A book by Henry Dreyfuss, *The Measure of Man: Human Factors in Design*, often becomes the sole human factors source material for an industrial designer.

Chief components of the ID curriculum may include such courses as drafting, strategies of material usage, development of sensitivity to product form, the use of light, shade and color in design, product aesthetics and subjective creativity, product forms for mass production, space analysis as a means of visual communication, industrial graphics, the use of two- and three-dimensional mockups, etc. Industrial design has been defined as the imaginative development of manufactured products and product systems that "serve the physical needs and satisfy the psychological desires of people."

It can be seen that ID curricula seem to present a broad, almost philosophical, premise in design. Use of such phrasing in curricula description as "humane performance justifying a product's existence" presents more of an abstraction than a technical basis for human factors. Indeed, a review of industrial design curricula in general tends to indicate the need for introducing a more fundamental human factors technology. The best means of presenting and assimilating this point of view for the ID student might be an integrative one. Thus, every phase of an industrial designer's training should ideally also incorporate valid human engineering principles. The abstract regard for "humanics" in industrial design must then be translated into meaningful perceptual-motor terms that can be realistically applied to design configurations.

Human Factors in Engineering

The diversity of puristic engineering fields also bespeaks a common need for incorporating the human factors technology into design. Aerospace engineering, agricultural engineering, ceramic engineering, chemical engineering, civil engineering, electrical and avionics engineering, mechanical, metallurgical and nuclear engineering may all require a well-integrated human factors component in the design curriculum.

Design is the common denominator of engineering. It is the iterative process through which an engineer is able to optimally convert available resources into devices or systems that satisfy the human needs to which the design is addressed. Engineering schools, of course, recognize that the engineer's discipline and skills are more fundamentally acquired through practice and application. Engineering students, they realize, lack the experience and knowledge necessary for evaluating optimal solutions or to perform the necessary detailed analysis to achieve these. They can, however, and they must begin at the outset to learn and to practice methods of the kind of orderly thinking into which their cumulative knowledge can eventually be integrated.

Engineering training programs emphatically assume the value and importance of designing for human needs; technical human factors data for design interfaces, however, are generally a neglected facet of engineering curricula. These should become an emphatic part of the information-gathering phase of design (Hammond, et al., 1979). In fact, in most engineering academic contexts, little appears to be known or taught about the human operator's makeup and perceptual-motor requirements. Though it would, perhaps, be neither necessary nor practical to require a special human factors curriculum component in an engineer's training, an integrated approach could be of considerable value for the engineer as well as the industrial designer. The information-gathering phase in any engineering project should also ideally incorporate valid human engineering principles to be applied in the design approach.

•

Material in the text has been organized to accommodate the curriculum needs of both industrial designers and the different engineering disciplines. It is expected that a measure of sound human engineering can be meaningfully integrated into any industrial design or engineering course or project. The nature of the human operator as an essential component in the design process is thus presented without going into any great depth. Indeed, more extensive human factors analyses can be left to the human factors specialist when necessary for the accomplishment of more complicated design solutions.

When human factors expertise is not immediately available, members of a design team can at least apply human factors analytical methods and design principles by knowing the rudiments of this science. Human factors approaches to analysis are thus described in the text. This, it is expected, will enable designers to apply meaningful substantive human factors data in design even in the absence of the specialist.

I
What You Should Know About the Human User

1
The Nature of the
Human Operator

To determine what is needed in the human factors aspects of design, it becomes necessary to understand certain fundamentals about the human user (HU) or operator. For purposes of design, the HU—or human operator (HO)—might be considered to be a kind of black box with a number of extrusions of body members as well as internal components (see Figure 1.1). It has a head with a brain, eyes, ears, nose and a mouth. It has limbs with muscles and a nervous network that connects to muscles, bones and internal organs. The organs keep the body fueled, operate in a combustion and waste-disposal process, and maintain equilibrium of the entire organism in balancing all the other subsystems.

The HO black box, for categorizing purposes, may be considered to be a complex system composed of four subsystems of body components:

1) The receptor subsystem, or the senses
2) The information-processing subsystem or the brain and the voluntary and automatic nerve networks,
3) The effector subsystem, or the muscles that hold and move the skeletal frame members,
4) The support subsystem which is made up of all the internal organs

The Receptor Subsystem
The receptor subsystem is made up of the senses, or channels, through which information from the external world enters the black box.

3

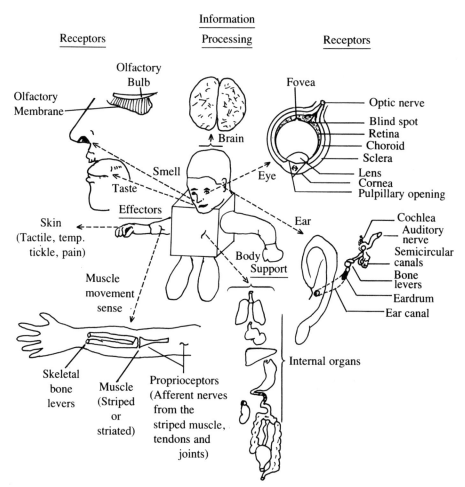

Figure 1.1. The human operator as a black box subsystem complex.

These are made up of the eyes, ears, taste, smell and the skin senses responsive to touch, temperature, tickling and pain. (The smell and taste senses are important in the chemical industry, and must also be considered when environmental annoyances and hazards are involved in design, such as in noxious gases and harmful liquids.) The semicircular canals are important to the sense of balance. The muscle-movement senses are called "proprioceptive" receptors; they tell us when our muscles are moving or what their static position is relative to the other muscle postures.

Ears

Hearing is a basic sense wherein signals are communicated from a distance. The total organism is alerted to outside events immediately before the touch senses become involved (even before sight). Hearing is physiologically a mechanical device that translates air vibrations into nerve impulses.

The external collecting mechanism is the outer ear attached to bone levers. The levers transmit the tiny vibrations to an "oval window" in the cochlea. The oval window is a membrane behind which is fluid. Air waves acting on the eardrum are thus leveraged via the bone levers, facilitating transmission to the liquid medium in which tiny hair cells are excited. The hair cells transmit the vibrations as nervous impulses to the brain.

Sound signals are efficient for gaining the HO's attention. However the sensing of signal direction is poor. Sound that occurs at both ears is most efficient for discriminating from the left or right; front or rear, and above or below directions, are often confused.

The semicircular canals in the inner ear contain fluid that stimulates hair cells when the head is rotated. These provide a sense of the upright position.

Eyes

The eyes provide the accurate sensing of light transmitted by objects in space. Eyes are sensitive to object motion in three dimensions, and to movement rates and the accelerations of objects. The eyes can discriminate thousands of shades and colors.

An eye is about 7/8 of an inch (22 mm.) in diameter. It consists of three coats—the sclera, the choroid and the retina—and is distended by internal fluids. The sclera is a white leathery membrane which becomes the transparent cornea at the front. The cornea is the bulging portion covered at front by a folded membrane called the conjunctiva. At the rear, the sclera is opened where the optic nerve enters the eye.

The choroid is a pigmented coating that forms a dark lining. At the front, it becomes a "ciliary body." The latter is joined with a muscle that accommodates the lens in focusing. The "iris" muscle regulates the size of the pupillary opening which gets bigger when light intensity is lowered, when drugs are used, or when a person becomes emotional.

Watery liquid fills an anterior chamber between the cornea and the lens. A jellylike substance fills the inside of the eye chamber, or the posterior portion of the eye behind the lens.

The lens is elastic and is attached by ligaments to the ciliary muscle. When this contracts, it pulls the choroid coat forward and the elastic lens bulges. Its increased curvature occurs in close-up focus. When the ciliary muscle relaxes, the internal pressure of the jelly-like substance in the eye chamber pushes the choroid coat back and flattens the lens again.

Retina

This is a photosensitive portion in the eye structure that contains microscopic rods and cones. The rods are denser at the forward areas of the retina. At the fovea, there are none. Cones are most dense at the fovea. Rods are sensitive to low-light levels or shades of gray, while cones are sensitive to daylight levels and color.

Dark and Light Adaptation

The eye responds to light in wavelengths from 400 to 760 millimicrons, where the sun reaches maximum radiant energy. It is least sensitive to the extreme ends of the visible spectrum. In the blue-green range, blue is seen at about 470 millimicrons, and green at 520. A blue-green color is seen within a mix of the spectral wavelengths between 470 and 520 millimicrons. The color red occurs at around 680 millimicrons, and yellow at 580 millimicrons. When light levels are low, red (at the high end of the spectrum) and violet (at the low end of the spectrum) become less visible than yellow. The rods are sensitive to these low-light levels. When fully sensitized, rods are 10,000 times more sensitive than when exposed to intense light. Rod adaptation occurs in the dark after about 20 or 30 minutes, but sensitivity continues to increase for up to 24 hours. After being exposed to the dark, the cones in foveal vision take about 90 seconds to adjust to light.

Afterimages

Since light produces a photochemical response in the eye lasting about a half minute, either positive or negative afterimages can occur. Positive afterimages occur after brief exposure; these are the same color and shade as the stimulus object. Negative afterimages occur after prolonged exposure. Thus, when a white profile on a black background is looked at for half a minute, the profile becomes black on a white background. Negative afterimages in complementary colors are also seen in such a fashion. When a blue image, for example, is fixated, a yellow one will be seen afterwards, or if red is fixated, a green afterimage is seen. In the latter case, the green smocks used in medical operating rooms are meant to blend with all the green afterimages occurring from excessive visual exposure to red blood, i.e., green is the complementary color to red because they occur opposite one another when arranged in a colorcircle.

Color Sensitivity

All colors can be seen when presented under bright light at the center of the visual field, which is called the fovea or macula lutea. All colors are seen as true when within 20 to 30 degrees of the center.

Color Blindness

Blue-green blindness occurs in about 4 percent of the male population. Only about .4 percent of females have this deficiency, where red, orange, yellow and green are all seen as yellow.

Contrast

This occurs when black is placed against white to set off one another. Blue against yellow and yellow against blue are more vivid than against other colors. Green is enhanced by red, which is known as "simultaneous contrast," because the afterimage of red is green. Such contrast effects can be employed by designers to enhance the sharpness of vision and to improve the attention-getting character of a display.

Binocular Vision

The disparity of an image that occurs in each eye, or binocular sensitivity, produces dimensional cues in depth perception. These slightly different angular views by the two eyes produce depth cues for up to 200 feet, or 61 meters.

Pattern Vision

This occurs in visual perception as a tendency to put forms together from what is seen. For example, reading can be understood as form perception since each word forms a configuration. Displays can often be arranged together to help the HO see forms or patterns more easily.

Smell Receptors

The olfactory organs are located in the recess inside the nose, and can be stimulated by sniffing in the scent. They are highly sensitive to minute quantities of volatile or vaporous chemical substances. The sensation of a specific odor, however, fades after only a few minutes, even though it may persist heavily in the air.[1]

Taste Receptors

The taste receptors are located on the tongue and the upper and back portions of the mouth. Only four tastes are distinguishable—salty, sour, bitter and sweet. Taste sensations combine with smell sensations.

Cutaneous Receptors—The Skin

The sense of touch serves in joining the meaning of sight and sound with body-contact experiences. For example, an object becomes identified as hot, cold, smooth or rough after the skin has contacted and joined with the visual experience.

Skin cells are stimulated through nerve endings that respond to pressure, pain, heat and cold, as well as the experience of movement on the skin, deep pressure, tickling, itching, and vibration. Pressure and pain sensing is in response to mechanical stimuli. Pain also occurs when any of the experiences are too intense on the senses.

The Information-Processing Subsystem

Nerve and brain cells are excited by the receptors or senses. The sensory signals are routed through a number of hierarchies in the nervous system, viz., the brain and spinal cord. The upper brain centers coordinate the sensory signals to produce effector responses. This coordination of signals is evident in a drunken person. When the higher brain centers are depressed from the effects of alcohol, a loss of control occurs, thus, exaggerated speech and poor coordination can be observed. Reflexes in such a person are not subject to the control of the higher brain centers,

---

1. Odors can be effectively used in design as alerting signals. They should, however, be as naturally alarming in character as possible, e.g., burnt, sickening or otherwise disturbing odors.

and so the head, eyes, trunk and limbs tend to pursue different courses of action.

The brain is the locus of the higher information-processing centers where overall activities are regulated. Special brain centers coordinate all incoming sensory signals for carrying out purposeful action by the effectors.

The Effector Subsystem

Effectors are the skeletal muscles that exert force to produce leverage in dealing with the external environment for regulatory and management action, such as for controlling tools, products and machines (see Figure 1.2).

Effective action depends upon how muscles are arranged around the skeletal system. In moving a bone about a joint, a muscle is attached at the place where the bone itself is moved to form a lever. Flexors bend a limb and extensors straighten it out, e.g., biceps and triceps of the upper arm. Muscle action either fixates a limb or throws it to and fro. "Synergistic" action occurs when bone groups in the body leverage system are fixated or held in place to accommodate coordinated movement of a distal or end member such as the finger or thumb. Movement in a finger-tracking task, for example, requires that muscles about the elbow and shoulder joints be fixated.

Skeletal muscles are attached to three types of joints that produce movement: (1) hinged, e.g., fingers, (2) pivot, e.g., elbows, and (3) ball and joint, e.g., shoulder and hip.

Types of Movement

Muscles hold a limb or body member in a fixed position, bend or extend the member, or execute rotary motion. It all depends upon how the muscle groups are arranged. Movement occurs as (1) slow, tense and antagonistic, or (2) ballistic, with brief pulses of muscle contraction that set a limb swinging freely without opposing action from antagonistic muscles.

Antagonistic Movement

Antagonistic muscle action is characterized by stationary postures or slow-moving fixations. In a slow-moving fixation, an opposing muscle

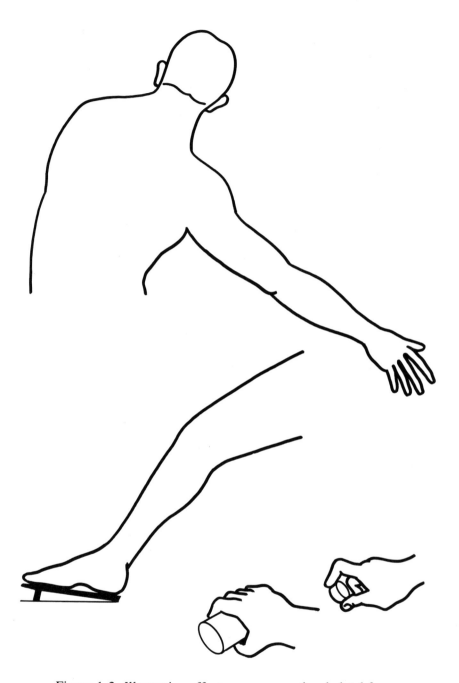

Figure 1.2. Illustrative effector organs on the skeletal frame.

yields to the contraction of its antagonist. The muscles act to coordinate a kind of slow pursuit movement, as in following a target with a control stick or carrying a glass of water.

Ballistic Movement

This consists of the alternate contraction of flexor and extensor muscles. The limb is thrown back and forth in this alternating muscle action. Ballistic action is exemplified by rapid tapping, piano playing, typing, simply pounding, etc. The upper limit of ballistic to-and-fro actions is around 10 per second. With finger action, however, a kind of "fusion" takes place, and the total number of possible strokes is increased to around 20 per second.

In learning complex muscle skills, ballistic action eventually emerges. An early learner is at first slow and halting; movements are tense and antagonistic. As learning progresses, antagonistic muscle action is reduced, and the slow, tense movements become the free-swinging ballistic ones.

Required Muscle Contraction

Muscle contraction in any activity can be described as either isometric or isotonic. Isometric contraction is required when a muscle must work against a constant force such as a spring. Isotonic contraction is simply a shortening of a muscle. Tension of a muscle can be measured through electrical changes. Isometric electrical changes are constant for a muscle group in alternating contractions. These changes in isotonic contraction are, by contrast, only momentary.

Isometric contractions yield superior performance since constant tension of the muscle results in continuing regulation of the action. On the other hand, isotonic contractions serve in regulating the task only while the contractions are in process, or while shortening of the muscles is occurring.

Fatigue

After repeated contraction, muscle fatigue results in: (1) a decrease in response speed, (2) a decline in strength, and (3) a slower return to the normal muscle length since the muscle remains partially contracted. Muscle fatigue occurs due to the depletion of stored nutrient materials and the production of waste products such as carbon dioxide, lactic acid and phosphates.

The Muscle-Movement Sense

The movement of muscles is due to nerve endings contained in the muscles, in their attachments to bone levers (i.e., the tendons), and in the joints themselves. The total sense of movement is called "proprioception," or feeling the action of all these components. The sense of movement itself is called "kinesthesis."

Adjunctive Motion Senses

Sensing the motion of the body through space, of course, is not a function alone of the muscle-movement sense. Rather, motion sensing is a complicated combination of signals from the vestibular balancing mechanism in the inner ear, dynamic changes in the visual scene as well as in sound cues, and in changes in skin pressures from accelerations at the seat and other various contact points. Tactual stimulation also occurs in movement when the skin is stimulated or stretched as a limb turns around its joint during muscle action.

The Sense of Equilibrium

The vestibular chamber contains tiny bones, fluid and hair cells that are stimulated with any angular motion or linear acceleration. This mechanism is an adjunct to the inner ear which contains both the auditory and labyrinthine membraneous subchambers. Canals within the subchambers are oriented for both vertical and horizontal stimulation. The equilibrium sensing thus provided, however, is sometimes unreliable in complex occurrences of stimulation. Spatial confusion and motion sickness can result from conflicting visual and vestibular signals, with the visual being the more reliable.

Reaction Time

The time from when a sense is stimulated until a response occurs in a muscle is called "reaction time." A portion of this time is taken up by the passage of the nerve impulse over the afferent or sensory nerve. Another portion of the time is taken up at the nerve center involved, and the remaining portion of the time is required for the passage of the impulse over the efferent or motor nerve and the explosive-like reaction of the muscle itself in responding. A simple reflex arc, involving only a minimum of nerve centers in the spine, takes about .04 seconds. When nerve centers in the brain or brain stem become involved, the shortest reaction time (RT) is about 0.2 seconds; e.g., an eyewink from a puff of air stimulating the eyeball takes about .04 seconds, whereas when the eyewink is "voluntary," it takes about 0.2 seconds since brain centers become involved. In addition, the greater the muscle mass that becomes involved, the longer the reaction time. For example, the leg and foot RT is much greater than that of the arm and hand.

Left-Handedness

HO efficiency can be compromised by inappropriate neuromuscular action requirements. Equipment and product designers, for example, frequently neglect hand-use preference in their designs. Left-handers make up approximately 15 percent of the total population and, when forced to use their right hands, they become less skillful and have longer RTs then their right-handed brethren.

The Support Subsystem

This subsystem brings nutrients and oxygen to the other subsystems and carries away waste products such as lactic acid and carbon dioxide. It is comprised of the heart, lungs, and circulatory and digestive internal organs. A nerve network supplies the body-support subsystem—the autonomic nervous system (ANS). It is arranged in clusters or ganglia of nerves located outside the spinal cord.

The ANS has two divisions—the sympathetic system and the cranial-sacral or parasympathetic system. The former serves in accelerating the heart, dilating the pupils, constricting surface blood vessels, stimulating the liver to produce blood sugar, dilating the bronchioles in the lungs, and shutting down digestive and sexual or genital functions. The cranial-sacral division is also known as the parasympathetic division since it functions generally opposite the sympathetic division to slow the heart rate, to constrict the pupils, to dilate the blood vessels, and to promote digestive and genital functions.

The ANS and the support subsystem are generally important for use as "criterion" measures for good design. For example, the metabolic rate, heart rate, respiration, etc. can all be used as measurements when evaluating the amount of human exertion required to operate equipment or to use products. These criteria of the support subsystem are most commonly used in industry for the development of work and performance standards.

Summary

In the understanding of human factors by engineers and industrial designers, the HO might be conveniently considered as a kind of black box. The problem then simply becomes one of understanding the nature of the black box and its characteristics in order to provide the best design accommodations for human operation.

The HO may be considered to be comprised of four subsystems as illustrated below:

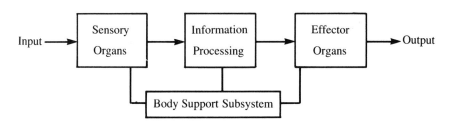

The senses of the eye, the ear, the skin and muscles, and the nose and mouth are most important to know in order to provide more optimal design features for products and equipment. The design most appropriate for human senses is readily perceived and understood. The nature of muscles attached to the skeletal frame should be considered by designers to provide the best, most easily controllable attributes of control devices.

II
Human Characteristics
Important to Design

2

The HO's Capabilities as Important Design Considerations

Product or equipment design requires that decisions be made about what the HO can and cannot do efficiently. Performance can be strategically enhanced if a system is properly designed, or it can be significantly degraded if the HO's capabilities and limitations are not adequately considered.

Basic human factors considerations for a designer must involve what the HO can best do when working with tools and machine products. HOs are typically employed for their sensitivity in sensing and reading signals put out by machine systems. The HO performs well when a machine is properly designed for making decisions and for performing certain manual tasks. The HO can surpass many machine functions because of a learning capability, particularly if this capability is adequately utilized by a designer.

How the HO Compares with a Machine Doing the Same Operation

In a tradeoff of capabilities when considering machine-design alternatives, the HO is generally superior to machines in sensory, information processing, and some manual control functions. The HO's special capabilities include:

1) *Perception within noise clutter.* The HO is able to filter out intelligible signals from complex visual and auditory noise fields, while machines are not yet able to do this.

2) *Multichannel perception.* The HO has various senses and sensitivities. This functional sensory diversity permits the HO to shift attention rapidly, which machines are also not yet able to do.

3) *Long-term memory storage.* The HO has excellent volume information storage from past experience. Medical diagnostic information, for example, as well the vast fund of information necessary for solving various design problems, would require inordinately costly hardware and software for a machine.

4) *Management control.* Integrative functions that an HO is capable of developing can provide an excellent judgment and decision-making control function. The HO is best qualified to take over when machines malfunction. Specific programs to handle specific malfunctions might not be anticipated by the programmers, while an HO on standby can still handle such management functions. The HO's inductive reasoning capability, wherein a condition can be sensed and generalizations made from it, also enhances management control. The HO, for example, can see that a robotic arm is stuck and needs a new pivot pin. A course of action can then be decided upon and shut it down and replace the pin. An HO might do this much more effectively than a machine. Machine strategies are limited to what can be preprogrammed in anticipation of such occurrences, and this is likely not feasible for all possible events.

5) *Versatility of input.* The HO has an effector mechanism capable of a wide range of actions. Intermittent and continual actions are readily performed by the HO, such as actuating switches or doing periodic or continual tracking. Such versatility also appears to be unaffected by such factors as age. This wide range of input potential is inherent in the HO's makeup, and providing a comparable capability in a machine would be costly.

6) *Better overall reliability in performance.* Machines are prone to rapid failure when breakdowns occur. The HO deteriorates in performance more gradually under conditions of work overload, fatigue or stress. The HO can thus hold up better than machines can for some functions.

7) *Enhancement of reliability.* Given time to work out solutions, the HO can provide a parallel reliability function that would be costly for a machine function to do. In fact, serial reliability is more generally the rule for machines. These two forms of reliability can be explained as follows:

Serial reliability $= C_1 \times C_2 \times \ldots C_n$

The reliability of each component (C) is multiplied to get the total reliability. Thus, if each component has a 99 percent reliability, a string of 100 such components in a piece of equipment would mean about one failure in three.

Parallel reliability $= \left[1 - (1 - r)^m \right]^n$
  where m equals the number of components,
    n is the number of functions, and
    r is the reliability of individual components.

Thus, if components are arranged as backup controls, they can each have only a low reliability but still give the unit good reliability. For example, if two parallel components have an individual reliability of only .90 each or fail one in ten times, then reliability equals

$$\left[1 - (1 - .90)^2 \right]^1 = .99$$

Less than one failure in 100 would occur even though each component has a low reliability. Employing the HO in parallel reliability design can thus significantly improve reliability.[1] Task-checking, for example, permits the HO to prevent a potential failure before it happens. When sufficient recovery time is available following a malfunction, the HO can, indeed, enhance the overall reliability of a system. Providing such a capability in a machine design—viz., by duplicating all components for parallel operation—might not be feasible from a practical point of view.

Typical HO Employment in a System

When interacting with a machine product, the HO is typically employed in detecting and reading display signals. Information processing functions are then carried out, followed by overt manual input. Display reading involves auditory, visual and other senses through which a ma-

---

1. For a further discussion of reliability and human factors, refer to W. Ireson's *Reliability Handbook* published by McGraw-Hill Book Company.

chine's signals can be read. Information processing involves closed-loop processes, as in deciding to throw a switch or in utilizing high-level decision-making capabilities through complex neural processes. The HO's effector system then executes manual control by responding in sequence or with the simultaneous time-sharing of functions.

Reading Signals

The HO senses or responds to display signals within a limited period of time when interacting with a machine. The signals may be intermittent or continuous, and the accuracy with which the HO reads signals can be measured by (1) the percentage of signals correctly detected, viz., the number detected divided by the total number presented, and (2) the number of false alarms divided by the total number presented.

Designing Display Signals

The auditory and visual modalities are most commonly used for displaying signals. Typical auditory signals used involve a tone to stand out in background noise. When designing such signals, their rate of presentation and the background noise must be considered. Typical studies of visual signals, often use displays such as a standard flashing light to act as a background noise stimulus. Intermittent brighter flashes serve as the signals to be detected.

Display detection can be improved under the following conditions:

1) *When the diversity of requirements is increased.* Tasks that are time-shared with signal detection, for example, should require the HO to shift attention and increase alertness. If a central control task is complex, such as in tracking or information-processing, the HO's reaction time can also be shortened. Thus, the HO's arousal level is heightened by the increasing task load.

2) *With the use of complex signals.* After practice, a group of subjects presented with complex signals surpassed the performance of a similar group presented with simplex signals. The complex-signal group was required to actuate a buzzer whenever the subjects heard any of the numbers "4,", "7," "8," "11," "14," or "18." Members of the simplex-signal group sounded the buzzer only when they heard the number "7." The simplex-signal group at first had fewer false alarms, but, with practice, the complex-signal group reduced their number of false alarms to fewer than those of the other group.

20

3) *When the signals are unambiguous.* Quality control inspectors required to identify only high-probability defects are most reliable. As much·as possible, display signals should be designed to signal only malfunctions or deviations that are clearly distinguishable.

4) *When effective training in signal detection is provided.* HOs should be trained under moderate levels of stress and given extensive practice on the signals to be detected.

Information-Processing

After signals are read, the HO typically processes the information. Information-processing occurs as HO judgments and decisions to be made. These are either simple internal responses, such as whether or not to stay within a speed limit, or complex decision-making responses, such as when to attack a military target. Complex decision-making involves subjective values that must be weighed against the probabilities of what might happen in a course of action. Individual risk-taking habits enter into such decisions to achieve a desired outcome. Some decision-makers accept higher risks in different situations than do others. Factors that are involved include:

1) The confidence that the decision-maker has in available information on outcomes.
2) The decision-maker's caution level.
3) The actual degree of risk involved.

The Decision Process

As a decision-making situation, consider a student's decision in selecting a career field. Subtle values of prestige and personal preferences must be involved in the decision to pursue a field of study. The risks here revolve about whether or not the values of the decision-maker will be truly realized. What is the likelihood that the student will sustain an interest in the field and all the individual subjects that it entails? What is the probability that an acceptable number of job opportunities will prevail after graduation? Will the same demand for services be there? Will the pay still be good? What is the risk that the student might be locked into a relatively low status throughout a career in the field? What risks is the student taking that he or she can handle all the course material successfully?

The decision process sequentially involves:

1) Assembling as much information as possible about the alternatives.
2) Analyzing courses of action and the possible outcomes.
3) Selecting the best course of action among the alternatives.

The diagram in Figure 2.1 illustrates the process of a college student selecting a major field as a career decision. The student first goes over the various fields of study with his advisor. Proceeding with the analysis, the student values certain conditions and weighs the likelihood of their being realized in the future. Is the field appealing and interesting? Are there job opportunities? Does the field pay well? What are the advancement opportunities? Then the student considers the relative difficulty of the curriculum. If the student can answer all questions positively, the selection is made, a curriculum planned, and initial courses taken. If at any of the decision points a negative case occurs, the student may repeat the decision process.

In a typical decision-making process, the subjective values of the decision-maker are weighed against their probabilities of achievement. The risk-taking propensity of the decision-maker is also involved in the final decision, of course.

Detailed probability information is more than likely lacking in most decisions. Objective probabilities for each eventuality are likely to be perceived by a decision-maker as to be nearly a certainty. Such perceived probabilities, however, often place operational decisions at risk simply because some of the variables are transient, such as a changing demand for workers in the selected field. The socio-economic probabilities in such transients are, of course, difficult to obtain. If accurate statistical data were available, these would most likely override the subjective values in the decision process.

Decision Games

A gaming approach to decision-making is frequently used in military war games. Business decision games involve such problems as production schedules and sales programming. In scheduling, for example, factory machines and personnel production capabilities are deployed to schedule orders in factories in the most efficient ways. Objectives in the simulation problem might include such things as minimizing delays, reducing production costs, turning out the orders on schedule, etc. Gam-

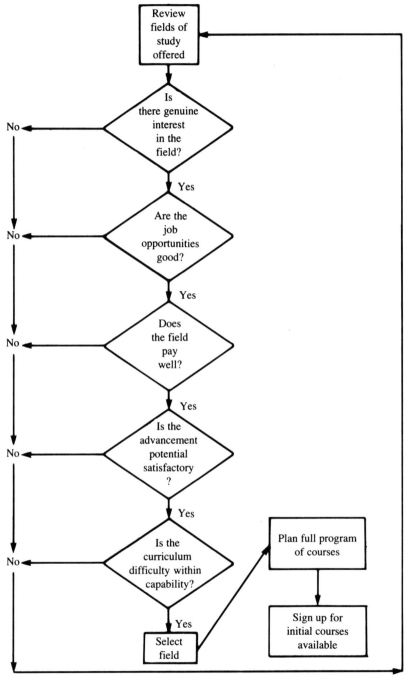

Figure 2.1. Decision diagram of a student selecting a field of study.

ing scores show the decision-game players how well they were able to meet the objectives in the face of competition. Studies show that such training is, indeed, of considerable value in acquiring decision-making skills.

Manual Control

In a machine model, display reading, information-processing or decision-making are followed by manual control. Manual control is typically carried out and characterized by the following:

1) An HO builds up a conceptual framework in visualizing the interactions with the machine to predict the consequences of control actions.

2) The HO must operate in a complicated way with a machine. Generally, practice alone is not sufficient to maximize a machine skill. The HO must be able to act in ways, and under diverse conditions, that are highly adaptive in the stress of emergencies and in learning the extremes of the machine response capabilities.

3) The tasks are all of a procedural or tracking nature. Procedural tasks involve sequencing discrete controls in positioning switches, pressing pushbuttons, moving levers, or turning rotary-position knobs. Tracking tasks largely involve following an auditory or visual signal in some response pattern. This involves such actions as steering an automobile and maintaining its speed, finding and pursuing a target with a marker on an oscilloscope display, aiming a gun, etc. Tracking tasks can be further categorized as "position" or "pursuit" tracking. The classical example of position tracking is shooting at a sitting duck. The pursuit tracking action would be shooting at the duck in flight. Most tracking tasks involve a combination of these two modes.

4) An HO's tracking responses are intermittent. Tracking intermittency or reaction time occurs at around 0.5 seconds. The HO's anticipatory behavior, however, can eliminate this delay time almost entirely. This is because the HO can see in advance where the target is going and responds to correct the expected error before it happens.

5) Manual controls have some degree of "feel." Feel can be incorporated by adding resistance or by springloading the control. (Friction loading may otherwise become necessary when vibrations and accelerations can effect the HO's response.) Springloading improves feel because of isometric effects that yield muscle cues. Force gradients, in-

creasing with deflection, incorporate velocity and rate cues since machine response increases with the amount of deflection. Pursuit tracking is also superior to position tracking due to the direction and velocity cues provided.

Control Feedback

A distinction should be made between sensory feedback and control feedback. Sensory feedback is comprised of machine signals returned to the HO from an operation. Control feedback refers to further control input to the system in its continuing operation. When sensory feedback is provided to the HO in tracking, it is termed "closing the control loop." Eliminating the sensory feedback is termed "opening the control loop."

The importance of sensory feedback for regulation in control feedback was demonstrated by a simple experiment. Subjects were required to move a stylus from a central position in front of them to a target area some distance from the starting point. Visual feedback was, of course, provided in the normal trials in which the subjects easily found the target. However, when the lights were turned off after they had begun to move the stylus toward the target, they generally missed it. Over short distances that were covered in less than 0.2 seconds, the loss of visual feedback made little difference since they were practically on the target to begin with. Longer distances requiring more than 0.2 seconds to be covered resulted in the target being missed altogether. Such a simple demonstration shows how important it is to see what it is you're aiming at. The loss of sensory feedback can thus effectively mean loss of control feedback as well.

Control feedback can be described as either "negative" or "positive." Negative feedback occurs when reversing the control for corrective action, while positive feedback reinforces or perpetuates the action, thus aggravating the error. When an HO turns on the hot water in a shower and the water begins to get too hot, the direction of the turn is then reversed to cool it. This is negative control feedback to keep the water from getting too hot. If the handle is accidentally turned to a hotter setting, the shower water gets even hotter and the HO may scramble out of the shower, perhaps with a yelp. The HO has thus reacted, in the latter instance, to produce positive control feedback. Normal thermostatic control is created by negative control feedback, i.e., the furnace is turned on when the temperature gets too cold, and off when it gets too hot. Positive control feedback would cause it to turn on when it's already too hot to get even hotter.

Another example of positive control feedback, that of creating instability, is in interpersonal conflict. When an angry response is met with another angry response, the situation can move toward violence. A calming response would thus be seen as negative control feedback.

Still another example is in the test flight of a missile system. Negative control feedback acts as a course correction to counter an error in the missile's direction. If erroneous signals within the system continue to steer the missile further off course, it can be said that positive control feedback is occurring, and when this happens in practice, the missile is destroyed to prevent it from crashing into a populated area. In this case, positive control feedback simply means that the system control continues to do what it is doing. Since, in this sense, positive control feedback fails to correct the error or malfunction in the system, it may be seen to be an inherently unstable control response, i.e., since the system continues out of balance until system control completely breaks down.

Manual Overload

An HO typically fills a multitask role when frequent, interspersed, time-shared tasks are called for. If the number of manual tasks an HO must perform approaches saturation, accuracy and efficiency of performance become marginal. When the workload first increases, the HO simply works faster and continues to handle it. When saturation occurs, some tasks are neglected, until eventually performance breaks down completely.

Increasing manual tasks to a marginal level may induce an HO to find more efficient ways of doing a job. With further workload buildup, however, performance becomes increasingly inefficient.

Work overload can sometimes be alleviated by such approaches as follows:

1) Careful selection of the most capable personnel.
2) Providing extensive task training.
3) Analyzing work sequences so as to shift task sequences.
4) Adding personnel.
5) Designing automated functions.

The HO's Learning Capacity

A major capability of an HO is the capacity to learn. The HO can be used in a system because he or she is not restricted to performing fixed, rigid operational sequences that are commonly identified with programmed or otherwise structured machines.

A fundamental principle of learning, however, is knowledge of results, sometimes called "information feedback." Educator Edward Thorndike demonstrated this principle with an experiment in which HOs were required to draw a straight line four inches long twenty times. One group was not permitted to look at their results. A second group placed a four-inch strip along each line to see how they did. A third group was told by an experimenter each time whether they were over or under and by how much. No improvement occurred with the first group where no knowledge of results was provided, but the other groups improved rapidly with the information they were given.

Knowledge of results has become a fundamental principle of learning. Indeed, to capitalize on the HO's learning capacity, it therefore becomes necessary to provide knowledge of results or information feedback for any design configuration.

Training Effectiveness

Learning experiences with which HOs are provided include three common types:

1) Classroom instruction.
2) On-the-job training.
3) Real-time simulation.

These training methods typically involve a procedure whereby a trainee advances to some designated skill level. This is based on a classical theory of learning that a trainee can only achieve some plateau and must advance beyond that on his or her own. Recent training techniques, however, have improved on this old theory and have incorporated the principle of adjustable task difficulty; this has been called "adaptive training." This technique provides for the problem or task to be changed as a function of how well the student is doing. In conventional training, the learning curve is relatively fixed. The student is taken to the plateau level without further advancement. In adaptive training, the student's skills are advanced by eliminating the plateau through adjustment of the difficulty. If the student is doing well, the difficulty is constantly increased; if he is doing poorly, it is decreased until he improves.

Adaptive training in Air Force training programs, for example, have used the following areas of difficulty:

- Stresses of the work environment, such as illumination, sound levels, temperatures, and gustiness or turbulence of air.
- Other environmental stresses of vibration, oxygen pressure, and drugs.
- Difficulty with control elements, such as gains and lags in control responses.
- Difficulty of problem in varying command trajectories that must be followed or maneuvers that must be performed.
- Secondary task-loading, such as increasing or decreasing the number of communication tasks, fluctuations of instruments, and the number of instruments that require reading and adjustment.

Thus, such adaptive training variables have served to improve the learning situation, making the learning curve linear throughout and providing a trainee with higher levels of manual skills.

Summary

Designers must first consider what the human user, for whom they are designing, can do with a tool or product. They must concern themselves with the kinds of capabilities the HO possesses in sensory functions, the way information is processed, and the way decisions are made. The nature of an HO's manual control response is important for a designer to know so that the best control configuration can be developed. The HO's unique capacity to learn must be considered by a designer, who can then determine how the user can best be instructed on what is needed to operate a machine product.

3
Designing for the HO's Limitations

In the previous chapter, an HO's inherent sensory and motor performance capabilities were pointed up as basic design considerations. Now,
the limitations of an HO will be discussed, since these are of equal importance in the development of safe and efficient design features. The
HO's sensory limitations and special restrictions in information-
processing and effector subsystem performance will be considered. The
effector subsystem will also be considered from the standpoint of bodily
dimensions (anthropometry) for the inclusion of essential spatial dimensions in design. The body support subsystem, with environmental
and other limitations on the HO's well-being and sound organic func-
tioning, will also be considered.

Sensory Limitations

Human senses are, of course, limited by the kinds and range of energy levels to which each sense responds. Sensory channels can easily
be subjected to overload, fatigue and injury, so the designer must protect them from any conditions that might impair their efficient function-
ing.

The hearing mechanism, for example, responds to vibrations through
a rapidly oscillating air, solid or liquid medium. The minimum detect-
able force of vibration is .001 dynes per square centimeter. The upper
limit of sensitivity is 1,000 dynes per square centimeter. Force levels
below the lower limits are either undetectable or are felt by other than

29

the auditory sense. Forces beyond the upper limit can be painful and can damage the hearing mechanism. The two-ear structure in the human body is also limited in sound directions that can be discriminated. Directions of sound from above or below, and front and rear, are most easily confused.

The visual sense responds to light in a spectral range of 400 to 760 millimicrons, which is only a tiny portion of the overall band of wavelengths. At the lower extreme of dark adaptation, the eye responds to levels as low as 1^{-10} footcandles. Upper extremes occur at around 10^4 beyond which pain and injury occur.

Skin response limits are two milligrams of pressure at the lowest level; high-level pressures compress the tissues and induce pain. Odors can be detected at very faint levels but fade after two to four minutes.

Temperature sensing is around 3° C. for cold and 45° C. for heat; beyond these limits, pain is felt to the tactile sense.

Information Processing

Sensory information is processed by the nervous system to produce an effector or skeletal muscle response. Some of the sensory information moves through quickly, as in a reflex arc. Other information acts to influence behavior through more complex processes before it has any effect on an effector response. Machine operations are generally handled at some intermediate levels of the nervous system. We do not, for example, think about everything that we hear and see. Sensory input from proprioceptive senses in the muscles, tendons and joints can produce muscle contractions without a person being conscious of them. This has some functional value in that it prevents confusion from information irrelevant to input requirements, and it makes for more precise information reaching the higher neural levels.

Processing Limitations

Actions of the higher processing centers of the brain can impose certain limitations on the machine processes by way of distractions, distortions and deviations. These, for example, might include the following:

1) *Stereotypes.* These are expectancies that operate in the information-processing subsystem that can distract, change or distort an out-

put. When the HO has learned to respond to a conventional mode of machine response, then that mode will persist in the future. Controls will be turned up for on, clockwise for increase, or, for that matter, for any way in which the HO has previously learned to respond.

2) *Objectivity.* The HO is a social creature. A machine, by contrast, will respond consistently without attitudinal influences. The HO will introduce social thinking about what "should be," about "good and bad," "right and wrong," or any other such value judgments which in turn will influence actions.

3) *Short-term memory.* The HO has a poor reliability for this capacity. If, for example, the HO is instructed to turn the heat off on the water after 15 minutes, it may or may not happen. A machine action must remind the HO to do this, or to do it itself.

4) *Information encoding and amplification.* Rapid categorization, assigning numbers to items, and summing types of activities are poorly executed by the HO. A machine can do these better. Otherwise, excessive training, practice and on-the-job experience are needed by the HO to translate, interpret and process the information before making decisions. The HO is also severely limited in the number of different operations that can be performed at the same time, particularly when high-speed processing is required.

5) *Emotional response.* The HO responds unreliably when involved in personal conflict, and reacts apprehensively about personal status. Performance is degraded as a function of neurotic disturbances, acute social tensions, and conflicts.

6) *Boredom.* The HO is a poor monitor of ongoing events when no action is required. The machine should be designed to monitor the HO to ensure that something is done when required, rather than the HO being simply held responsible. Thus, flashing lights, auditory signals, etc. should be designed to alert the HO when action is required. The HO's performance also falls off when long duty cycles and repetitious tasks are involved.

7) *Stress reaction.* Information-processing becomes unreliable when the HO experiences severe stress. While mild forms of stress can facilitate performance, excessive stress can gravely compromise it. Stress can be intensely disruptive when a threatening situation develops that places an HO's life or job in jeopardy. In such situations, behavior becomes rigid, stereotyped and inappropriate. Under severe stress, performance breaks down and the HO becomes an unreliable systems component.

Effector Subsystem Limitations

In providing for efficiency of the effector subsystem, a designer must consider the different sizes of people in the user population, lifting and force-exertion capabilities, and reaction-time and tracking-performance limitations.

Anthropometry

A number of different anthropometric data sources are available. Various bodily data have been extensively compiled on military populations. Nonmilitary anthropometric data have been compiled by the Public Health Service, the National Bureau of Standards, and a number of commercial firms. The Bureau of Standards has compiled data on children and infants for garment-sizing. The Public Health survey has compiled perhaps the most extensive civilian anthropometric data on adults, ages 18 through 79 years of age. These were noninstitutionalized populations where height, weight and 16 other dimensions were measured.

Commercial firms have also developed anthropometric data over a number of years for civilian clients. The industrial design firm of Henry Dreyfuss Associates has compiled data for human-scale models taken over a 35-year period (Niels, et al., 1974 and 1980). Another design firm in Cupertino, California, Design Edge, has prepared "ergomanikins" for use by designers based on a range of body sizes from the 5th percentile female to the 95th percentile male dimensions.[1]

Figure 3.1 presents some common baby and children anthropometric dimensions prepared by the National Bureau of Standards (NBS). The NBS has since discontinued these compilations since they are now being performed by the American Society of Testing and Materials. The data in the figure represent composites of male and female statistical averages. These are employed as standards since the differences between the sexes for the age groups studied were considered negligible (National Bureau of Standards, 1953).

Civilian anthropometric data for adults have been somewhat restricted due to the limited availability of subjects in random sampling. This is not the case with military populations, however. Figure 3.2 presents an-

---

1. Insomuch as statistical measures are extensively employed for anthropometry and other human factors data, some statistical sense of measurement is important for designers. Appendix A presents a brief description of the meaning of statistics for designers.

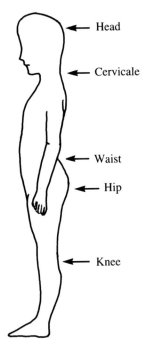

Figure 3.1. Anthropometric data for babies and children—
composite measures (from Commercial Standard CS 151-50,
Bureau of Standards, 1953).

AGE IN MONTHS (APPROXIMATE)

	3	6	12	18	37	63	88
Height (in inches)*	24	26.5	29	31.5	37	43	48
Weight (pounds)	13	18	22	26	34	44	54
Head to cervicale (head and neck length)	5.7	6	6.2	6.4	7	7.5	8
Cervicale height	18	20.5	22.8	25.2	30	35.5	40
Cervicale to knee	13	14.5	16.2	17.5	20.5	24	27
Waist to knee	6.8	7.7	8.7	9.7	12	14.5	16.5
Waist to hip	2.6	2.8	3.2	3.5	4.4	5.2	5.7
Crotch height	7.7	9.2	10.5	11.8	14.8	18.4	21.4
Knee height	5.2	6	6.7	7.6	9.5	11.5	13

* Divide by .3937 to obtain centimeters.

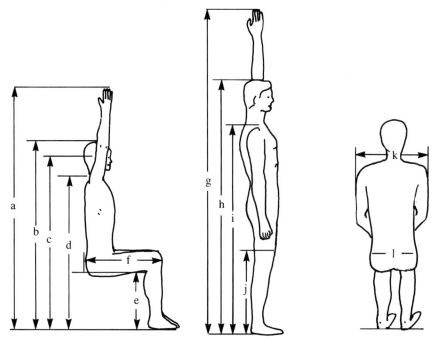

Figure 3.2. Common anthropometric measurements
(U.S. Army population).

Percentiles (in inches*)

	Males			Females		
	5th	**95th**	**1st to 99th**	**5th**	**95th**	**1st to 99th**
Sitting vertical arm reach (in inches)	50.4	58.1	48.9 to 60.2			
Sitting height	33.0	38.1	32.0 to 39.2	30.9	35.0	30.0 to 35.9
Eye sitting height	28.3	33.2	27.4 to 34.4			
Midshoulder sitting height	22.3	26.8	21.2 to 27.9			
Popliteal height	16.1	19.2	15.5 to 19.8			
Buttock-knee length	21.7	25.6	20.9 to 26.6	20.5	24.3	19.7 to 25.3
Standing vertical reach	82.0	93.1	80.2 to 95.9			
Standing height	64.3	73.0	62.8 to 74.5	60.0	67.8	58.5 to 69.5
Shoulder height	52.8	60.8	51.2 to 62.4	49.3	56.4	47.9 to 58.1
Knee height	18.8	23.1	18.1 to 24.1			
Shoulder breadth	17.0	20.3	16.5 to 21.6	14.1	17.4	13.5 to 18.4
Hip breadth sitting	13.2	16.7	12.6 to 17.5	12.9	16.5	12.3 to 17.7

* Divide by .3937 to obtain centimeters.

34

thropometric data obtained in a military setting. Military data are generally the most comprehensive since subjects can be readily obtained in drawing large representative samples. Military data may be considered to represent overall percentiles of a healthy U.S. adult population up to around age 44.[2] In fact, when civilian data are not available, the anthropometric data derived from a military population may be quite acceptable. For example, when comparisons are made between civilian anthropometric data (Stoudt, et al., 1965) and USAF trainee data (Department of Defense, 1981), close similarities can be observed. Consider the following comparisons of military trainee data with those of comparable civilian populations in the age range between 18 and 24.[3]

The civilian population percentiles, except for weight, appear to be quite comparable. The use of the more extensively researched and comprehensive military anthropometric data is feasible when generalizing to similar age groups in the general civilian population. Such data can be most useful when seeking special measurements such as facial and hand dimensions that have not been made available in the anthropometry of the general population.

Longitudinal studies of military anthropometric data have indicated little change over time. Recent generations do not seem appreciably to

			Percentiles			
Male	1st		50th		99th	
Height (inches)	62.6*	(62.9)†	68.6	(68.8)	74.8	(74.8)
Weight (pounds)‡	115	(111.1)	157	(148)	231	(220.7)
Female						
Height	58.4	(59.2)	63.9	(63.9)	69.3	(69.8)
Weight	91	(122.1)	126	(122.1)	218	(162)

* Civilian
† USAF Trainees
‡ Divide by 2.205 to obtain kilograms.

---

2. Extremes at the distant tails of a distribution—i.e., 99.999 percentile or 00.111—of course would not apply.
3. While it is true that the military has some body-size restrictions in their recruitment standards, these data show that the 1st and 99th percentiles are almost identical to those of a civilian population (Stoudt, 1965). The extremely small and large body sizes found in civilian populations would, of course, not be found in a military population, but these would be less than two percent of the total civilian population.

differ in body size from the earlier generations when age is held constant (White, 1979).

Measurements of the over-65 population differ from those of younger populations in the 19 to 44 age groups in the following ways:

1) Males become shorter with age, also evidencing a loss in weight.

2) Female stature declines with age, but weight increases.

3) Males fall off in height by about three percent. An elderly male's average weight is also around 18 percent less than that of a younger adult male.

4) Elderly females lose five percent of their average height, but gain 10 percent in weight. Over age 75, their average weight also falls off.

Such changes in body measurements for the elderly may be due to a number of physiological and structural developments:

- Bone loss.
- Weakening of the extensor muscles in holding an erect posture.
- Thinning of the cartilaginous discs between the vertebrae.
- Thinning of the vertebrae themselves, producing an outward spinal curvature that reduces stature.

The decline occurs primarily in sitting height. Dimensions that do not involve the spine, such as leg and arm lengths, remain about the same as in younger populations.

Anthropometric Design Strategies

A number of different approaches can be taken in accommodating human body sizes in design. The designer, for example, may fit a tool to his own hand size, or a shelf to his standing height. Such an approach is more random than meaningful, however.

When considering the overall body sizes of a consumer or user population, more practical approaches include designing for a median body size, or at the extremes for, say, the 5th and 95th percentile body sizes, or for an adjustable fit for a full range of population body sizes.

Designing for the Median

This also means designing for the average or mean size of a user population. In this strategy, a design compromise is made in using an average size. Cabinets, tables and other furniture, for example, might be simply too costly to design for a complete range of user sizes from the extremely small to the extremely large. This design inconveniences the larger and smaller than average sizes, but a large percentage would be accommodated in the middle range, i.e., at the hump of the normal bell-shaped curve.[4]

Designing for Extremes

This design strategy is commonly used. The upper and lower limits are accommodated; thus, the height of a door designed for the largest body size would accommodate all smaller sizes. Knee clearances at a console, designed for the largest buttock-to-knee dimension, would accommodate all smaller such dimensions. When the lower extremes can be accommodated in reaching and grasping controls and in seeing all displays, then people who are larger should also be able to reach and grasp controls or to see the displays.

An Adjustability Strategy

This is, perhaps, the best design approach and should be used whenever possible. It simply provides for an adjustable range of body sizes, such as seat height, helmet diameters, etc. In military equipment design, the range between the 5th and 95th percentiles is generally used. This avoids overly-costly adjustments while providing for the most convenient use by 90 percent of the population. Extreme body sizes, in this 90-percent approach, cannot be accommodated and are generally rejected for use in the system.

A Full-Range Strategy

Designing for the accommodation of the complete range of population body sizes may sometimes be possible and desirable. In fact, such a design strategy may be particularly pertinent where survival operations are involved, as in the case of handholds and outside dimensions of escape chutes and hatches. In other cases the costs might just not be that much more for providing a few additional inches of adjustability or clearance, thereby making the product accessible to everyone in the population regardless of size. These are considerations that are incumbent upon the designer to make when assessing the utility of his design.

---

4. Modular design approaches could also be considered in accommodating a greater range of body sizes, where cost factors may not be prohibitively more than those of a single standard design.

Biomechanical Limitations

The HO's lifting and force-application ability is limited by muscle, bone and tendon leverage structures of the body. Pivotal body members, their articulation and synergistic tension points in various activities, are illustrated in Figure 3.3 (Grieve, et al., 1975). Lifting and force capability is a function of torquing forces that can be applied at various leverage points of the body.

Synergistic Muscle Action

Synergistic action occurs when body members proximal or closer to the center line of the skeletal frame must be held fixed so that those closer to the tool or work can do the manipulation. In lifting an object, for example, straining torque is applied to the back or lower-lumbar spine and to articulation points centered about the wrists, elbows, shoulders, pelvis, knees and ankles. Lifting force can be most effectively applied:

1) When the bulk of the load is within one foot of the body.
2) When the load is at a height several inches above floor level.

A straight-back, bent-knees method of lifting is recommended to facilitate lifting. Lifting capability is hampered:

1) When the required height of the lift increases. Lift capability is lowest at shoulder level.
2) When the horizontal distance from the body increases. The capability is least at full arm length.

Figure 3.3 shows some of the points at which synergistic muscle action is required to hold body members in position in order to execute work action. In shoveling, for example, the straining torques are applied to the upper and lower back, the shoulders, the upper and lower arm, and to the wrist. In addition, knee-bending, posture-holding tension torques may also be required in the shoveling posture with conventional shovels.

Conventional handtools such as pliers, screwdrivers, etc. require a bent-wrist posture as illustrated in sketch A of Figure 3.3. Redesigned tools, such as in sketch B, can relieve some of the synergistic tension. Bent-arm postures required in control design (sketch C) and bent-leg postures (sketch D) introduce further tension torques. Effective tool and machine design can do much to alleviate such straining torques, i.e., de-

38

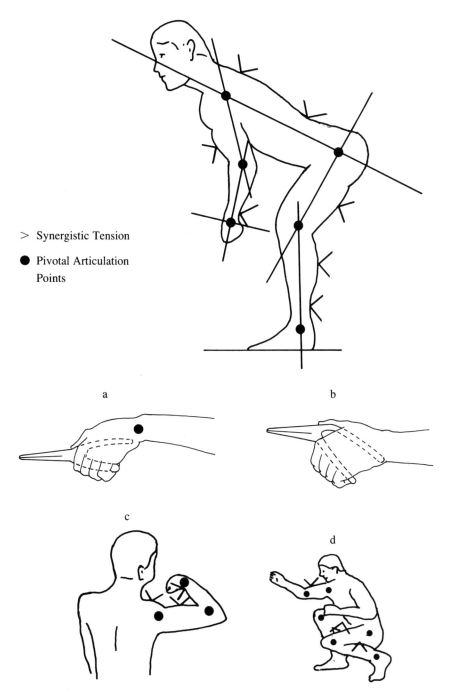

> Synergistic Tension

● Pivotal Articulation
  Points

Figure 3.3. Synergistic tension and pivotal articulation in body action.

signing the bends into the tools rather than requiring further synergistic muscle actions, locating controls at proper levels to eliminate synergistic holding postures, etc.

The 5th percentile force capabilities in various postures for a male are outlined in Table 3.1. Studies of female force capabilities indicate upper-body strength to be about one-third that of males, with lower-body strength about one-half. Myoelectric studies have also indicated force-exertion capability by males to be two or three times that of females (Kroemer and Marras, 1981). Strength tests taken of U.S. Army women have shown that an average peak lifting force is around 140

Table 3.1. 5th percentile force exertion of U.S. Army males.

Posture	Force Exertion
Lifting when within one foot of the body	85 pounds* up one foot†
	35 pounds up 5 feet
Carrying	65 pounds for less than 5 yards
Lever force, right arm at center of body:	
Pull	50 pounds (24 pounds when down and right of center)
Push	50 pounds (24 pounds when down and right of center)
Raise up	9 pounds
Lower (press down)	13 pounds
Move to left	13 pounds
Move to right	8 pounds
Gripping, momentary	56 pounds
Gripping, sustained	33 pounds
Gripping with thumb and forefinger only:	
Momentary	13 pounds
Sustained	8 pounds
Leg pushing force, when upper-lower leg angle is 150 degrees and back is supported	500 pounds

* Divide by 2.205 to obtain kilograms.
† Divide by 3.281 to obtain meters.

pounds with a lever 15 inches above the floor and when using arms, shoulders and legs with the back straight. Men are able to lift half again as much. The 5th percentile capability for women with this task is about 90 pounds, and for the 95th percentile, 200 pounds (Hertzberg, 1972; Tichauer, 1978).

Lifting capability is largely limited by the torque applied to the lumbar spine or the lower back. Low-back injuries occur most frequently when load design places the hands at wide horizontal distances from the body (Figure 3.4). Close-in loads are lifted primarily by the powerful leg muscles, with only a back-supporting function necessary. Loads that are designed out away from the body require actual lifting by the lower back itself. This applies a high torque force on the lower back or lumbar spine, frequently resulting in back strain and injuries. Work studies have also indicated that musculo-skeletal injuries may occur when such heavy loads are repeatedly lifted.

Response Speed

An HO's response is limited by the maximum possible speed of human reaction. As mentioned before, minimum reaction time, for all practical purposes, is from 0.2 to 0.3 seconds and then only when expectancy is operating, i.e., the HO anticipates a signal and is ready to respond. In reacting unexpectedly, reaction time is increased to 0.5 seconds or more.

The HO's tracking speed in pursuit tracking is generally limited to three radians (171.9 degrees) per second.

Load close
to the body

Load out from
the body

Figure 3.4.

HO Limitations Due to the Environment

The HO is limited in the use of products or in operating equipment by the different conditions that occur in the ambient or outside environment, and by the conditions caused by operation of the machine itself. These latter conditions occur from machine-generated noise, fumes exuded, etc. These general conditions may include such parameters as noise and vibration, radiation, various conditions of light, various gases, temperature and humidity, motion and acceleration.

Noise and Vibration

The types of noise are:

1) Impulse. This occurs simply as abrupt, periodic changes in the sound pressure level yielding different frequencies of disturbance. Examples are a jackhammer or someone moving furniture. Its annoyance value occurs as a function of intensity and duration.

2) Steady state. This class of noise is characterized by its constancy, such as the steady level in a factory or on a busy street. Its level and frequency can be measured in terms of:

• potential hearing impairment
• levels of discomfort
• Speech interference
• Performance degradation.

Vibrations affect the HO at low frequency levels and relatively high amplitude.[5] Intense low-frequency vibrations acting on the body can interfere with manual and perceptual performance as follows:

1) Fatigue and nervousness occur during prolonged exposure.

2) Physiological disturbances. At a vibrating frequency of 5 cycles per second, the large body organs resonate. The vibrations can occur in either the X, Y or Z axis, or all of the axes; they can be random, sinusoidal, or both; and their frequencies can be up to 25 cycles per second with amplitudes of from .12 to .20 inches to occur as disturbing vibrations.

---

5. Infrasound, i.e., below 20 hertz and/or of extremely low amplitude, may occur only as a vague sensation. At greater amplitude, it is manifested as shaking or physical vibrations.

3) Vision is impaired in the frequency range from 10 to 25 cycles per second.

4) Manual performance is most impaired at frequencies less than 5 cycles per second. Only reaction time and monitoring behavior do not appear to be affected by vibrations (Grether, 1971).

Radiation

Radiant energy is classified by wavelength and frequency, viz., radio, microwave, infrared, visible, ultraviolet, X-ray, beta and gamma rays. Dosage levels are measured in units called Roentgen Equivalent Man (REMs), and the most harmful radiations are X-rays and gamma rays. Radiation effects depend upon the length of exposure and the part of the body that is irradiated. The Federal Radiation Council limits total-body exposure to 3 rems for any calendar quarter, and 5 (N − 18) rems for the total lifetime of an individual, where N equals the present age of the HO in years.

Light Conditions

The upper level of light for functional vision is around 10,000 foot-candles. Visual functions, such as acuity, contrast effects, and the ability to discriminate different intensity levels asymptotes or approaches zero improvement at around 100 foot-candles. Rod sensitivity, or the ability to detect low-light levels, diminishes after exposure to bright light. After exposure to 1,000 foot-candles, for example, it takes 20 minutes in the dark to detect a level of .0001 foot-candles. After exposure to light intensity for an instant above a million foot-candles, one minute or more is needed to recover vision for seeing even in bright light.

Since vision is a photochemical action, its constancy is due to the rapid microscopic shifting of an image on the retina. If the image is caused to be fixed in the same spot on the retina—i.e., the shifting of the image has been prevented—it will fade and become a temporary blind spot.

When a light source is intermittent, a "flicker" or a blinking effect will occur. If the frequency of the intermittence increases to around 30 to 40 times per second, depending upon light intensity, the flicker will disappear. Flicker should be avoided since it can result in visual discomfort and, in some individuals, can induce brain dysrhythmia.[6]

Discomfort can also be caused by glare. Glare can be either simultaneous or successive, as in going from dark to bright illumination.

Principles of good lighting should encompass visual limitations as follow:

1) Provide high visual contrast when speed and accuracy in reading are required, e.g., black on white, yellow on black, etc.

2) Avoid blue-colored illuminants when sharp vision is needed.

3) Provide illumination levels consistent with task requirements.

4) Provide controls for glare:

- Avoid light sources lower than 60° from the line of sight, and provide shading when necessary.
- Use indirect lighting to avoid point light sources.
- Use multiple small lights rather than one big one.
- Use nonreflective surfaces for floors, equipment and work surfaces with less than 30 percent reflectance values.

5) Provide relatively steady light when possible. Prolonged exposure to the frequency oscillation of fluorescent light, for example, can result in a kind of intermittent aftereffect of the stimulus called "buzz vision."

6) Use complementary colors with care. They can be enhancing when in proper proportion and intensity. However, a "flash" effect can occur if they are improperly used due to eye muscle tension induced from attempting to focus on the fluctuating complementary color effect. This can become somewhat physically disturbing when an HO is unaware of the condition that causes it.

---

6. Brain dysrhythmia can be induced in susceptible individuals, and can result in disturbances ranging from "peculiar feelings" to "flicker fits" and epileptic seizures. As much as one-fourth of the population may experience such sensitivity to a greater or lesser degree. The flicker stimuli can be an erratic fluorescent light, rotating helicopter blades, highway posts flashing by, or a psychedelic light show. Red flicker may be especially disturbing, with females often more sensitive to flicker than males (refer to M. Ferguson's *The Brain Revolution,* New York: Taplinger Publishing Company, p. 75 ff.). Most disturbing flicker rates are those approaching the normal alpha rhythm of the brain, around 10 to 12 cycles per second.

Gases

Normal atmospheric gases are necessary in permitting an HO to freely take in oxygen and expire carbon dioxide and to maintain normal gas pressures within the body.

A normal atmosphere is composed of 79 percent nitrogen and argon, and 21 percent oxygen with a trace of carbon dioxide. The pressure of the atmosphere at sea level is 14.7 pounds per square inch (psi).[7] At 20,000 feet, the pressure is reduced to 10.1 psi, and at 25,000 feet it is 5.4 psi. Reduced atmospheric pressure can result in diminished energy from a gaseous exchange in the lungs and excessive nitrogen pressure ("bubbles") painfully expanding in the limb joints. The latter is called "the bends."

In the vacuum of outer space, blood would boil; internal gas pressures would literally explode in the tissues. Air surrounding the HO must thus be kept within the limits of internal body pressures. High oxygen levels must be provided at low ambient pressures to permit efficient oxygen exchange in the lungs.

Contaminants

Relatively low levels of air pollution can be tolerated to maintain HO health and safety. Air pollution is measured in parts per million (ppm). The most common machine-generated pollutants include carbon monoxide, ammonia, nitrogen oxides and aldehydes. Many of the other common industrial pollutants are listed in Appendix B. Heavy doses of these pollutants can cause organ damage, and can often be lethal.

Carbon monoxide, as a combustion product, is odorless, tasteless and colorless. It has a high affinity for red blood cells, thus displacing oxygen in the blood. This results in disorientation and mental collapse. Effects are cumulative. Concentrations that can be tolerated safely for short times can become dangerous and fatal after long exposures. The limit is 50 ppm.

Nitrogen oxides are also not easily detected, but their effects can be fatal. They are dangerous because little or no discomfort is experienced

---

7. Multiply by 70.31 to obtain grams per square centimeter.

while they are being inhaled. Yet, after 12 to 24 hours, they can be fatal. The limit is 5 ppm.

Ammonia can result in eye irritation, unconsciousness and death. The limit is 50 ppm.

Aldehydes are also among the waste products of combustion. They can cause eye irritation and nausea.

Temperature and Humidity

Performance efficiency is limited to a relatively small range of temperatures. Surface contact temperatures above 95° F.[8] require skin protection. Air temperatures above 85° F. can deteriorate performance. The preferable temperature levels are:

- Summer with light clothing—between 70° and 80° F.
- Winter with heavier clothing—between 65° and 75° F.

Relative humidity should be between 30 and 75 percent. Low relative humidity is necessary for comfort at higher temperatures, but it should never go below 15 percent so as to prevent skin drying.

Motion and Acceleration

Motion effects can disorient and nauseate an HO. The semicircular canals, normally employed in spatial orientation as the chief sensory mechanism, can become unreliable during rotary acceleration.

Acceleration is the rate at which motion or velocity is increased. It is defined in terms of Earth's gravitational force, called simply "g," and is a propelling force of 32 feet per second every second, i.e., every second the velocity is increased another 32 feet per second.[9]

High levels of acceleration degrade performance and subject the HO to discomfort and possible dangerous physiological effects. The body planes to which accelerations can be applied are head-to-foot, called "positive accelerations," foot-to-head, called "negative accelerations," and front-to-back, called "transverse accelerations." In positive accelerations, blood is pushed toward the feet, taking blood from the brain and causing possible "blackout." This happens when 3 or more gs are ex-

---

8. Multiply by 5/9 and subtract 32 to obtain centigrade.
9. Multiply by .3848 to obtain meters per second.

erted for as little as perhaps 10 seconds. Blood is pushed up toward the head in negative gs, resulting in hemorrhages ("redout") in the brain. This occurs at 2 to 3 gs after only a few seconds exposure. There is considerably more tolerance to transverse gs, which can be tolerated briefly at 12 to 15 gs. However, the HO is still subjected to impaired breathing and labored movements.

For any significant accelerations, trunk and limb supports are necessary. Other control aids are also necessary to prevent loss of control, such as straps to hold in the limbs, friction-loading of the controls, etc. Physiological protective devices are also necessary to alleviate harmful acceleration effects.

Design Limitations of Children and Adolescents

In addition to physical size dimensions when designing for children and adolescents, designers must also consider their unique limitations in sensorimotor skills, information-processing, and the internal support subsystem.

Almost all males and females have achieved their adult weight and height by around age 20. In the baby, the head and body are largest, and the legs are short. The adolescent is slim and gangling. His face changes strikingly during adolescence; it becomes longer, and the nose and mouth grow larger. The internal organs increase markedly in size and weight during puberty. By three years of age, most children can climb and walk. At four, they can climb stairs, and by five they can jump, skip, hop and dance.

The "maturation" process pretty much restricts a child's motor skills. A group of children 24 to 36 months of age, for example, when given extensive training in buttoning their clothes and cutting with scissors, perform no better than a lesser-trained group, indicating that the neuromuscular system cannot be hurried.

Performance gains in tapping rates have been found to double (from around 80 to 160 taps per 30-second period), and grip strength to increase fivefold (from around 10 to 50 pounds of force[10]) from the ages of 6 to 18.[11]

---

10. Divide by 2.205 to obtain kilograms.

11. The average pull strength for children under 10 years of age is around 8 pounds, as compared with a pull-force capability of greater than 70 pounds for a 20-year old.

The strength of male and female children is about the same until around age 12, after which males continue to surpass the female in body strength well into adulthood. On average, boys also tend to excel in speed and coordination of gross bodily movements. Observations of pre-school children have indicated that boys are faster and make fewer errors in walking along narrow boards. They are also more accurate in throwing a ball, and can throw it farther.

Girls generally excel in manual dexterity. They are usually able to dress themselves earlier than boys, and evidence superior finger and wrist movements, such as in washing their hands, turning doorknobs, buttoning clothes, and tying bowknots.

Females are also superior in color discrimination, with color blindness eight times more common among males than females. Female superiority in color discrimination has been found as early as infancy. In taste, smell, hearing and touch, no significant differences in discriminability have generally been indicated between the sexes.

Such physical limitations and characteristics as these must be studied and applied in detail when designing for children and adolescents.

Limitations of the Aged and the Impaired

Designers must be sensitive to the special limitations involved when developing products for the aged and the impaired. The aged, for example, evidence decrements in the receptor and effector subsystem capabilities, as well as in such support subsystem components as the cardiac and respiratory organs (Bromley, 1966; Schonfield, 1974; Fozard and Popkin, 1978; Hughes and Neer, 1981; Welford, 1981). In designing hand-operated instruments such as nutcrackers, as a case in point, the grip strength for breaking nuts of an 80-year-old female can be expected to have diminished by 15 percent from that of a 20-year-old (Bromley, 1966). Likewise, the average speed of response and upper-body strength fall off with age. Visual acuity, hearing and smell sensitivity also decrease, while glare sensitivity increases.[12] Such limitations are, of course, all important design considerations.

Products must be given special consideration when designed for peo-

---

12. While auto accident rates decline with age, this is largely due to a decrease in driving activity. A rate increase, rather, is indicated when mileage is factored in for the over-65 driving population.

ple with sensory, motor, information-processing and stamina impairments. Sensorimotor impairment relates to such debilities as the blind, the deaf, the spastic, the arthritic, wearers of upper- and lower-limb prosthetic devices, etc. (Nickerson, 1978; Sheridan and Mann, 1978). The mentally retarded can be generally considered to be the information-processing impaired, which may necessitate the application of a design rule of simplicity (Wade and Gold, 1978).

Summary

Sensory limitations in hearing, seeing, feeling, taste and smell should be taken into account in design, as well as the characteristic limitations involved in processing information derived from the senses. The HO is a social creature, subject to stereotyped reaction, poor objectivity, emotionality, boredom, and deteriorated performance under stress. Limitations of the effector subsystem should also be considered in design relative to various body sizes or anthropometric limitations, and the strength and biomechanical structures of the body in lifting, manipulating tools, and other work activities.

Limitations imposed on HO performance—as a result of the outside environment or as a result of conditions generated by machine or industrial processes—should also be included as design considerations from the standpoint of the HO's health, safety and performance.

The limitations cited in performance, and those due to the operating environment, apply when designing for all HOs. In addition, design considerations must be given to the special limitations of children and adolescents, and of the aged and impaired.

Exercise 1—Taking Anthropometric Measurements.

Obtain your own anthropometric measurements as follows:

1) Height
2) Sitting height, erect
3) Sitting height, normal
4) Knee height
5) Popliteal height
6) Elbowrest height

7) Thigh clearance height
8) Buttock-knee length
9) Buttock-popliteal length
10) Elbow-to-elbow breadth
11) Seat breadth
12) Hand breadth
13) Hand length

If you have no special measuring equipment, be as accurate as you can using yardsticks, tape measures, rulers, etc. Team up with someone to obtain the measurements. Obtain your percentiles for the dimensions measured using military or public health population data. If you use the Dreyfuss or Niels data, indicate the populations from which samples were drawn.

It may be necessary to interpolate or approximate your actual percentiles since only interval percentiles are often given in reference data. Show the 5th, 50th and 95th percentiles from the population you use, as well as your own percentiles.

References

Department of Defense. "Military Handbook of Anthropometry of U.S. Military Personnel." Washington, D.C.: DOD HDBK 743, 1980.

Niels, D., et al. *Human Scale*. Volumes 1 through 9, Cambridge, Mass.: MIT Press, 1981.

Stoudt, H., et al. "Weight, Height and Selected Body Dimensions of Adults in *Vital and Health Statistics*, Series 11, No. 8, Rockville, Md., Public Health Service, 1965.

III
Human Factors
Design Data

4

The Application of
Human Factors Design Data

Human factors principles can be applied to products or equipment systems at virtually any level of complexity:

- Simple handtools—hammers, screwdrivers, pliers, snowshovels, hand spades, toothbrushes, etc.
- Simplex machines—electric hairdryers, electric drills, sanding machines; appliances such as toasters, clothes dryers, clothes washers; radios and tape decks; gasoline and electrical lawnmowers, etc.
- Complex machines—automobiles and trucks, diggers, cranes, earth-moving shovels, etc.
- Complex systems—mass transit transportation, computer robotics, ground and air weapons, spaceflight vehicles, etc.

Applying meaningful human factors methods and principles to design projects may sometimes require independent considerations from other aspects of design. Designers may be primarily concerned with how well equipment functions or how a product's appearance might affect is marketability, but such emphases may not necessarily include important human factors considerations. Nuclear power plants, for example, have large duplicated control-display consoles that are often laid out in a mirror-image fashion. HO position habits acquired at one console are then transferred in reverse when working at the mirror-image console whose controls and displays are laid out in directly opposite locations. Such design techniques may have significance to other aspects of design or may simply be design expediencies, but, in any case, they do not serve the HO well in facilitating good performance.

To assure relative proficiency in the HO's performance, the designer must apply the proper human engineering methods and design principles. Any approach to the development of human factors design criteria must thus include at least the following:

1) Identification of all functional interactions of the HO with equipment or system products.

2) Analysis of control-display requirements.

3) Application of pertinent human engineering data, either derived from the current state-of-the-art or from special design studies.

4) Determination of the full gamut of conditions under which operations will be carried out. This includes both the natural environment and the machine-generated environment.

Control requirements must be determined, together with the associated displays required for feedback. Populations of users should be identified to determine demographic and physical characteristics important to design. Health and safety considerations require that a detailed analysis be made of the conditions to which HOs will be exposed and the protective provisions that will be required.

The entire design can then be HO-oriented so that understandable nomenclatures can be used (significant to the users and not simply an engineering expediency), and the specific capabilities and limitations of the users can be considered and incorporated as design criteria.

Packaging and Handtool Design

In designing packages and unpowered handtools, be they snowshovels, screwdrivers, hammers, etc., basic biomechanical principles of design should be considered. Whenever possible, synergistic tension should be minimized. In packaging-lifting design, for example, lifting handles and the weight center should be located close to the HO in the horizontal plane to prevent lower-back strain. Handtool design should basically consider principles of good grip and body geometry. When feasible, several functions should be combined into one tool, such as the nail-pulling claws on the back side of the hammer. This approach to design can save time for the user and reduce fatigue.

Grip Surface

Three types of gripping design might be considered: power, precision, and general control (Emanuel, et al., 1980). In a power grip, the thumb and finger interlock. In a precision grip, the thumb and finger manipulate the tool for delicate positioning. The maximum diameter, in this case, should be .25 inches, with a soft rubber or wooden surface for finger traction, i.e., rather than metal. For general control, such as in simple one-hand positioning or two-hand manipulation, a maximum of a three-inch diameter should be provided.

Geometry

When considering the biomechanics of tools, the bend should be in the tool design rather than in the hand-wrist, arm, back or legs. Bent wrists, for example, not only produce synergistic tension in flexor muscles, but can result in "tenosynovitis." This results when tendon friction occurs at the wrist bone. Such tension can be avoided in the design of pliers, brooms, hammers, shovels, etc. by incorporating a slightly bent handle design, referred to as the "Bennett design." A double curve in opposing surfaces of the handle can also improve handling in conforming to the contours of the grasping hand. A finely jagged (knurled) or cross-hatched (serrated) handle surface can also improve handling when hands are wet or slippery.

Simplex Machines

Many of the principles in human engineering that apply to the design of handtools can also be applied to the design of simplex machines such as clock radios, electric can openers, washer-dryers, etc. In addition, human factors principles of control-display and console design can also be applied to simplex machines. Such tool design principles as good grip surfaces can also be applied here. A hair dryer grip surface, for example, can be serrated to prevent its being dropped when the hand is slippery. Indicator lights on radios and digital alarms can be appropriately color-coded for simplex machines just as for more complex machines.

Complex Machines

Human factors principles may be totally applied to complex machines where control-display and console design requirements should be spelled out in detail. Chapters 5 and 6 will treat such requirements. In addition, several detailed human factors standards and guidelines can be found in the literature to assist in a design application. These include:

Department of Defense. "Human Engineering Design Criteria for Military Systems." These criteria are equally applicable to civilian systems.

Eastman Kodak Company, Human Factors Section. *Ergonomics Design for People at Work.*

T. Kvalseth's *Ergonomics of Workstation Design.*

C. Morgan's *Human Engineering Guide to Equipment Design.*

H. VanCott's and R. Kinkade's *Human Engineering Guide to Equipment Design.*

W. Woodson's *Human Factors Design Handbook.*

W. Woodson's and D. Conover's *Human Engineering Guide for Equipment Designers.*

Complex Systems

A full gamut of analytical methods and human factors principles can be applied whenever total system complexes are involved in design. This means that every aspect of a system operation is considered—the complete matrix of equipment stations, the interstices, the support and maintenance aspects, etc. The value of applying human factors principles throughout a complete system is in utilizing the human factors technology to its fullest; the complete range of human operator capabilities and limitations can be considered in every facet of the system. This should permit the designers to come up with a more optimal design for humans to achieve total system goals.

The power of the system approach for optimizing human factors resides in being able to manage the technology in a totally flexible way. For example, reducing the automobile accident rate in a national program could be vastly facilitated by focusing on those subsystems that are most responsible. The traffic-control subsystem might be redesigned to reduce the accident rate, rather than simply the vehicle subsystem which is only partially to blame.

In all cases, the detailed human factors that are involved can be prominantly exposed and appropriate principles can be effectively applied to optimize the design.

Summary

Human factors principles in design can be effectively applied at all levels of complexity, from simple handtools to complicated systems of equipment. Careful analytical approaches are necessary to assure that the HOs and user populations are identified by their characteristics, and that appropriate human factors considerations are made. In the design of handtools and in the packaging of products, appropriate biomechanical principles should be incorporated into design. Simplex machines should incorporate the human factors principles that are not only appropriate to the biomechanics of handtools, but to the broader considerations of control-display design that pertain to more complex machines. The full gamut of human factors methods and design principles can be applied when a total system complex is involved. The power of such an approach lies in the potential for optimizing all human factors aspects of a system so that it can accomplish overall operational goals.

5
Control-Display
Design Principles

Control-display design principles have been formulated over a period of
many years on the basis of both field and laboratory experience. If these
principles are not applied in design, it, of course, does not mean that a
product is necessarily doomed to failure, or that a system is sure to
break down. It does, however, mean that human operators can operate
more efficiently in the system, and that lower human-error risks can be
achieved when these principles are appropriately incorporated into the
design.

Controls

Controls, from a designer's standpoint, are devices through which the
HO can make an input to a product, piece of equipment or machine system in regulating operations. The devices can involve a simple switch
operation or can be as complex as a three-dimensional tracking stick.
Complex interdependencies can also operate between and among the devices. It must be remembered that the way in which the devices are designed will basically determine how efficiently the HO will be able to
perform the regulatory functions.

Displays

Displays are any of a variety of mechanical, electrical or chemical
intelligence-producing apparatuses designed to stimulate the HO's

59

senses in order that the control devices can be appropriately regulated. Displays can be designed as visual, auditory, cutaneous (skin), olfactory (smell) or gustatory (taste) apparatuses to detect and read a status, condition, position, movement or direction, thus enabling the HO to respond appropriately with control devices.

Control-Display Design Criteria

Appropriate control-display configurations in design can only be based on a perspective of what the HO must do, and what he or she must know in order to do it. Traditionally, an HO performs such control functions as:

- A discrete or single-action control, such as in turning on a system.
- Position selection or setting a control on one of a number of fixed positions.
- Selecting a quantity at a variable scale for a given numerical condition.
- A tracking operation for a continuous period in regulating a course. This is exemplified by a unidimensional task, such as in left-right steering control or longitudinal speed control. Multidimensional tracking involves two or more dimensions in simultaneous control, such as in flying an airplane.

Figures 5.1 through 5.4 illustrate a number of control types and their characteristics as used in executing control actions. Others means of control action, using body members other than the hand and foot, have also been devised. A pilot's helmet, for example, called the "Darth Vader" in the operation of military aircraft, permits pilots to aim weapons at targets through head action. This was developed at Wright Patterson Air Force Base in Ohio. Versions of the helmet have been developed for the Army's Apache helicopters and the Navy's F-4J jets. Such helmets have been operational for several years, with versions depicted in such movies as "Blue Thunder" and the television series "Airwolf."

In the Army and Navy models, a pilot spots a target and flips a switch. An electronic system senses the position of the helmet and aims missiles or guns in the direction the pilot is looking. When a target is seen as the head moves in the helmet, the sensors register the coordinates and the missiles are launched or the weapons are fired. The HO

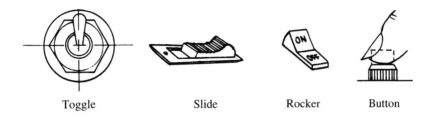

Toggle   Slide   Rocker   Button

PULL LEVERS

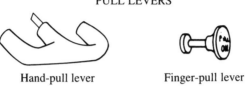

Hand-pull lever   Finger-pull lever

Figure 5.1. Discrete-action controls.

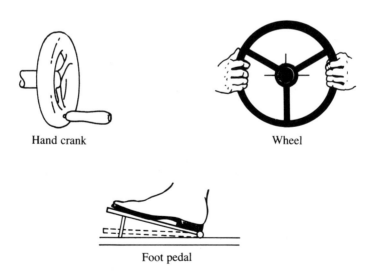

Hand crank   Wheel

Foot pedal

Figure 5.2. Unidimensional tracking controls.

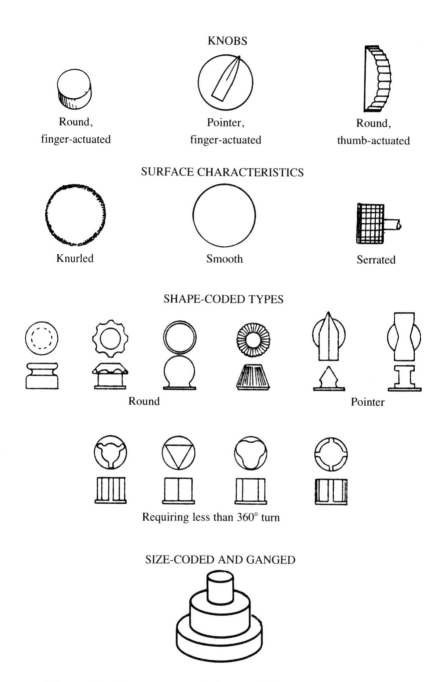

Figure 5.3. Some types and characteristics of knob controls.

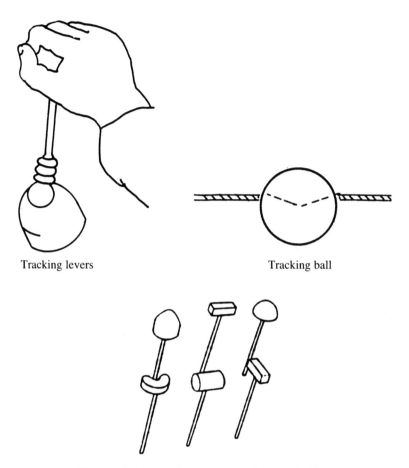

Tracking levers   Tracking ball

Shape-coded control knobs for tactual discriminability

Figure 5.4. Multidirectional tracking controls.

can aim his head many times faster than when tracking with hand-held controls. Such newly-devised control devices as these may thus have very considerable advantages over the more traditional hand-foot means of control.

The types of information necessary for executing any of the control functions can be described as follows:

1. Check displays or discrete signals indicating the presence or absence of a condition.

63

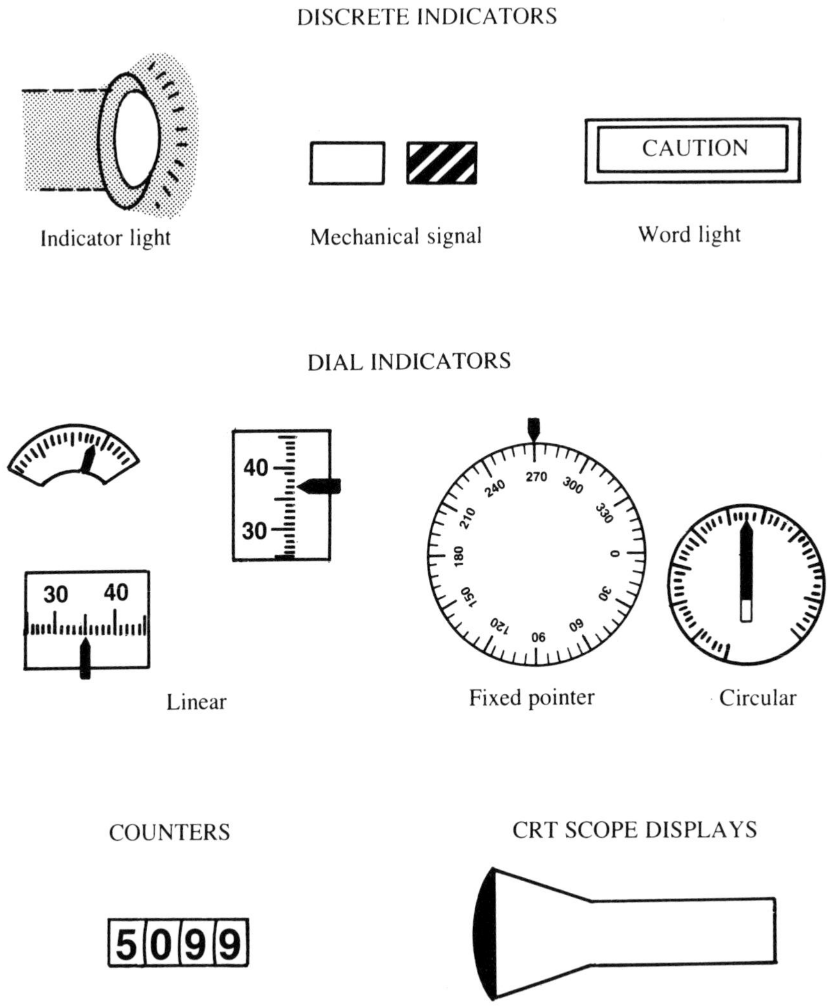

DISCRETE INDICATORS

Indicator light   Mechanical signal   Word light

DIAL INDICATORS

Linear   Fixed pointer   Circular

COUNTERS   CRT SCOPE DISPLAYS

Figure 5.5. Some display types and characteristics.

2. Qualitative displays indicating movement and direction relative to a condition or control action.

3. Quantitative displays indicating a specific quantity or numerical measure of a condition or control function.

In general, check displays include such items as indicator lights, mechanical flags and other visual or auditory displays signalling a go-no go

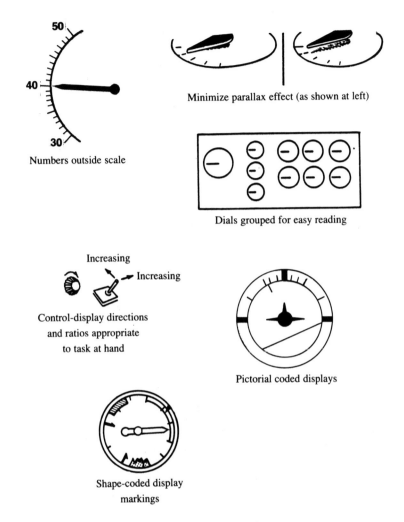

Numbers outside scale

Minimize parallax effect (as shown at left)

Dials grouped for easy reading

Control-display directions
and ratios appropriate
to task at hand

Pictorial coded displays

Shape-coded display
markings

Figure 5.6. Some preferred display characteristics.

condition. (Other discrete signals can, of course, also be used, such as cutaneous-vibratory or pressure stimuli, etc.) Linear dial displays showing pointer deflection, or cathode-ray tube (CRT) scopes showing targets and their deflections with placement markers, are generally considered best for qualitative displays. Counters and other easily-read numerical displays and digital readouts are best for quantitative readings. Figures 5.5 and 5.6 illustrate several types of these displays.

Control Selection Criteria

Control selection criteria include:

1) The panel space required for hand or finger clearances that are required for actuation.

2) The accuracy and force required in manipulating the control. Such considerations should determine the configuration to be employed and for what body member the control should be designed. (Hand-fingers control provides the best accuracy; leg-foot control provides the greatest force capability but the slowest reaction time.)

3) A need for nonvisual locations for controls. Blind reaching, for example, may be required when attention is directed elsewhere. Thus, location, shape, size and mode-of-operation[1] coding should be provided for easy nonvisual discrimination. Coding for rapid and easy visual access should include color and pictorial techniques.

4) The need for the control to show emphatic position settings. This can be accomplished by switch displacements or pointer knobs and other visual configurations showing unequivocal (e.g., nonparallactic) position settings.

Table 5.1 presents an analytical exercise in trading off one control against another to select the best design application. These, of course, are only examples of the more extensive human factors criteria that can be applied in tradeoff studies. More direct questions to be asked in selecting controls involve:

1) What are the elements to be controlled and how do they affect total operations?

2) How precise must the control be?

3) Is the operation normal or of an emergency nature?

4) How fast must the HO reaction be?

5) How will feedback information in control response be provided?

6) What are the forces to be overcome by the control? What is the range of its movement and how fast must it travel?

7) What information about the control does the HO need, such as the setting status (i.e., its position), the direction it moves with respect to its display, etc?

---

1. Mode-of-operation coding is provided simply by different ways of operating adjacent controls, such as by means of "click," different directions for actuation, and other ways of differentiating the feel among the different controls.

8) What are the elements or objects controlled and how are they moved? What is their direction and how fast must they be moved?

9) What is the required rate or ratio of control movement to that of the display?[2]

10) How should the control be grouped with other controls according to operation sequences, logical similarities, etc.?

11) What should the control size be? For example, what clearances are necessary for hand operation where a large hand is about four inches wide and a finger about one inch, and more where gloves are required?

A number of control-design rules might include the following:

1) The number of controls should be minimal to reduce the required frequency of reaching and manipulation actions. For example, two or more controls can sometimes be combined, such as one switch that turns several different functions on and off at the same time.

2) The control should be located adjacent to its display (below or to the left for left-hand operation, and below or to the right for right-hand operation. This prevents the hands from blocking the display.)

3) Controls should be easily identified by labels indicating what is controlled, e.g., "FLOW RATE." Directions for on-off, increase-decrease, etc. should also be labeled.

4) Labels should be brief. Only standard abbreviations should be used.

5) Labels should be located consistently, i.e., either always below or always above, but never mixed to prevent confusion.

Coding Controls

Control-coding should include both visual and nonvisual methods. Visual methods involve such design techniques as coloring controls or bracketing them with outlines, using visual shapes and sizes to identify them, etc. Nonvisual methods should involve techniques wherein the HO can discriminate the controls by touch and feel or by relative location. Coding techniques can, of course, be used in combination for both visual and nonvisual operations. For example, a large red T-shaped handle is often used for emergency situations. This is both visually and nonvisually discriminable.

---

2. This is called the "C-D ratio," and is simply the amount of control movement compared to the amount of movement in the display element. A large amount of display-element movement for a little amount of control action is called "slewing."

Table 5.1. Illustrative criteria for control selection.

| Criteria | CONTROL AND RATING | | | |
	Poor	Fair	Good	Excellent
Must show single position setting	Pushbutton	Slideswitch	Rocker switch	Toggle switch displaced 30°
Must show multiple position settings	Array of pushbuttons	Knotched slideswitch	Round rotary knob with marker	Detented rotary pointer knob
Must take up only a minimum of panel space	Toggle switch	Rockerswitch	Slideswitch	Pushbutton
Must be inaccessible to accidental actuation	Toggle switch	Rockerswitch	Slideswitch	Recessed pushbutton

Must provide some grasping surface	Pushbutton	Rockerswitch	Slideswitch	Toggle switch with ¼-inch knurled tip
Must provide for quick-ness of response	Foot pushbutton	Toggle switch	Rockerswitch	Fingertip pushbutton
Must provide for accurate scale setting	Momentary toggle switch	Knurled thumb-operated knob	Pointer knob	Serrated 1½-inch round knob
Must provide for high force exertion	Tracking ball	Handwheel	Vertical tracking lever	Foot-actuated horizontal bar with back brace
Must provide for efficient target tracking in two or more directions	Two or more control knobs	Tracking ball	Pressure-sensitive vertical control lever	Displacement-sensitive, springloaded vertical control lever

Color-Coding

Color-coding controls is effective when the color meaning is significant, such as the use of red for emergency controls. Color schemes should be used consistently. Blue, for example, should be used only when that same function occurs again in the system. Appropriate color standards should also be used throughout the system, such as red for emergency, green for safety, yellow for standby, etc. When sharpness of vision or acuity is an important consideration, along with the need for identification in coding, colors should be selected at the long wavelength end of the spectrum. Good acuity should be possible, for example, with the use of yellows and reds. A consistent finding in experimental studies is that illumination from the short end of the spectrum —such as the color blue—is inferior in acuity to that of other monochromatic wavelengths or to white light (Baker and Grether, 1963).

Shape-Coding

Only shapes that are readily distinguishable by touch alone should be used. Shape-coding is of value in low-level lighting or when vision is directed elsewhere.

Size-Coding

The number of controls size-coded for touch discrimination is limited to three sizes when the HO cannot feel all of them to discriminate by size alone. Shape- and size-coding can be used together to enhance discriminability.

Control Grouping

Related controls should be placed together and outlined by a border, with meaningful labels for the groups.

Control-Motion Stereotypes

Expectancies should be designed into controls.[3] Some control actions are commonly understood stereotypes, such as turning a switch up for

---

3. Though control-display stereotypes may largely be learned expectancies, an inherent psychological predisposition seems to operate in certain instances. A stereotyped response seems to operate under a wide variety of experimental conditions

on, pushing a throttle forward to increase power, or moving a wheel to the right to turn to the right. However, control-motion stereotypes may not always be so clear. A designer should then complete a simple study to determine the stereotype by using a procedure such as follows:

1) Set up the control situation, but do not label the control direction.

2) Using a sample of typical HOs, explain what the control does.

3) On cue, have the subjects move the control after being told to move a vehicle, a crosshair or other display element in a given direction.

4) Determine statistically what the significant motion stereotype is. Using several subjects, this can be measured as percentage differences in the control directions moved, i.e., when significantly higher in one direction than the other, this is the stereotype.

Display-Selection Criteria

Displays may be categorized by visual and auditory types. Auditory displays have been rather limited in design applications until recently, and voice-generation technology has been advanced in numerous computer product applications. However, the technology is still competing with that of visual displays. Specific applications may require careful tradeoff studies.

Table 5.2 presents an analytical approach in applying selection criteria to the design of both visual and auditory displays. As in Table 5.1, these examples are only meant to illustrate the approach for how a more extensive array of display design criteria should be considered in tradeoff studies when selecting displays for specific applications.

Visual Displays

In the application or use of visual displays in design, the following rules may apply:

1) Only that information necessary to execute control should be provided. Otherwise, excessive clutter results.

---

wherein there is an expectation for a display element to move in the same direction that the control moves, i.e., the HO identifies with a display element and tends to move it in the same direction that he or she would move.

Table 5.2. Illustrative criteria for display selection.

Criteria	Poor	Fair	DISPLAY AND RATING Good	Excellent
Must get HO's attention quickly	Moving dial at pointer	Indicator light	Flashing indicator light	Auditory signal
Must tell the HO what is wrong	General auditory signal	Labeled indicator light	Word light	Voice information signal
Must provide qualitative or trend information	Counter	Dial with fixed pointer	Circular dial with moving pointer	Linear dial with moving pointer
Must provide quantitative or numerical information	Fixed pointer on linear scale	Fixed pointer on circular scale	Moving pointer on circular scale	Counter or digital readout
Must permit precise numerical setting	Discrete light or mechanical signal	Fixed pointer on circular scale	Moving pointer on circular scale	Counter or digital readout
Must provide for signal tracking in two or more dimensions.	Two or more indicator lights	Two or more circular dials	Two or more linear dials	Cathode ray tube (CRT) scope

72

2) Design should conform to common reading habits or expectancies and stereotypes, i.e., increasing clockwise, forward, up or to the right; consistent with reading conventions progressing from left to right, top to bottom, etc.

3) Display elements should have good contrast, such as black letters and numerals on a white background.

4) Display pointers should have minimum parallax and should be unambiguous where the pointer end is clearly distinguishable from its other end.

5) Trade names and information not immediately pertinent to the reading task should be deleted.

6) The HO should be immediately alerted to displays critical to operation, i.e., placed within the 30-degree cone of vision, flashing, etc.

Shared Displays

When displays are to be shared on a console, the reading distance between the console and the different HOs assigned must be taken into account. The size or diameter of a display (with proportionately sized markings, of course) can be approximated according to the required viewing distance as illustrated below.

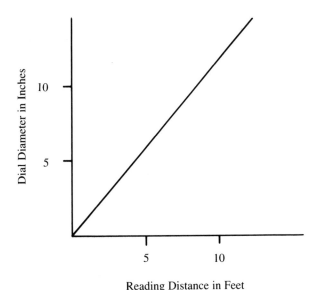

Reading Distance in Feet

The number of large shared displays, however, should be minimized since they tend to equivocate responsibility. Indeed, shared displays can only be justified on the basis of task/work module assignments given the individual HOs. Provisions for large group-read displays tend to compromise reliability since no single HO may be assigned to be completely in charge of the given functions displayed. Thus, an HO may come to expect that other HOs are or were or will be in control of the functions. When the question of, "Who should do it?" is in any way left hanging, then no one may emphatically assume control; thus, the responsibility is equivocated and the reliability of performance is compromised.

The Use of Colors in Display Design

The use of color in displays can be profligate in a kind of aesthetic profusion, or it can be most effective in enhancing the HO's efficiency. Studies have shown that mismatching words and colors can retard the HO's reading time (Schiebe, et al., 1967). When the word "grass," for example, was presented in green, reading time was reduced. When presented in blue or red, reading time was significantly increased.

In a study of the association of word meanings with specific colors, strong connections were found to occur between the word and color (Warren, 1974). When asked to identify colors with descriptive adjectives, connections were made as follows:

Red	Yellow	Green	White
"stop"	"minor"	"active"	"clear"
"alarm"	"standby"	"enable"	
"critical"		"manual"	
"failure"		"on"	
"major"		"run"	
"disable"			

In the past, the improper use of color light signals has resulted in a large measure of confusion for the HO (Parsons, et al., 1978). Red lights, for example, are often used in nuclear power plants to indicate an on or flow condition. Green lights have been used for an off or on-flow condition. When the color of an indicator light is not selected for a commonly understood meaning, as cited in the Warren study above, it can create confusion which can and often does result in human error. Rules for color-coding displays and indicator lights should be applied in design as follows:

1) Only standardized colors should be used, viz., red for emergencies, yellow or amber for malfunctions or a noncritical nature, etc.

2) The number of indicator lights should be reduced to an absolute minimum to avoid HO confusion. A "Christmas tree" effect can often become dangerous in critical operations.[4]

The Impact of Color on Mood, Emotion and Design Appeal

A widely held principle in display design is that colors can be used to affect moods and design appeal. Room colors in mental hospitals, for example, have been found to influence patient emotions, e.g., red-colored rooms for cheering up depressed patients, blue-colored rooms for quieting excited ones, etc. (Birren, 1965). The use of different colors in facility or product design can thus be of considerable importance. In this sense, colors are used for sensory-environmental effects as well as for display-signal intelligence.

Research on the effects of color has involved physical and subjective aspects. A prevailing view, for example, has been that there are thermal effects in the use of red for warmth and blue for coolness. The evidence has indicated that perceived relationships exist for the use of proper colors on both a physiological and intellectual basis (Eysenck, 1941; Berry, 1961; Acting and Kuller, 1972; Bennett and Rey, 1972). Hospital rooms colored in white are judged to be more "open" than those colored in green. Green-colored rooms are considered more "complicated" or "motley" than those colored in white.

Children have been found to deem rooms colored in light blue, yellow-green, or orange to be more "beautiful" than those colored in white, black or brown. Their behavior and performance is positively effected when they are placed in what they consider to be the "beautiful" rooms.

So, a thorough color analysis can be of major importance for creating the desired environmental impact in design.

The use of colors in packaging is also indicated to have a major im-

---

4. At the Three Mile Island thermonuclear power plant incident, during the critical malfunction over 100 indicator lights came on. The indicators most significant in identifying what went wrong were lost in a proliferation of irrelevant signals.

pact on the user. Female consumers, for example, in appraising the strength of the same soap powder in different containers, described the soap when packaged in a yellow container splashed with blue as "too strong." The same soap, when packaged in a blue container splashed with yellow, was judged to be of a more moderate and acceptable strength.

Labels

Control-display labels can be either worded or pictorial. In each case, they should unequivocally indicate what the display reads, what the control does, or what the HO should do. Here are principles of good labeling:

1) Labels should be consistently located.

2) They should be readily seen and not persistently blocked by the hand or a structural member of the compartment.

3) Only common abbreviations and symbols should be used.

4) Lettering should be unembellished and easy to read. Old English script, for example, is aesthetic in appearance but difficult to comprehend.

5) Superordinate large labels and subordinate smaller letters should be used for identifying groups and their subgroups.

6) The smallest characters should not be less than 1/8-inch in height for viewing at a 28-inch viewing distance.

7) Colors should be employed in labels only when illuminated by white light.

8) Labeling should be augmented by spatial separation, visual shaping, and other types of coding to enhance readability.

Signs

Signs are most efficient when they can be rapidly read and easily understood. Performance criteria for signs, therefore must involve accuracy and speed. Some conditions for reading speed and accuracy are:

1) Illumination levels should be a minimum of 25 millilamberts of white light.[5]

---

5. A millilambert is approximately the same as one foot-candle, which is the illumination provided by a candle on white paper from a distance of one foot.

76

2) An unembellished character style should be used to minimize reading difficulty.

3) The preferred size of characters is a function of illumination, viewing distance, and approach speed. By day, a 10-inch letter size should be adequate for reading from 75 yards away, for example.

4) Proper colors can improve visibility. International orange is generally seen at the greatest distance. Red on white, or bright yellow on black, and black or blue on bright yellow are good against a wide variety of background colors. Red or orange markings are readily seen against sky and foliage.

5) Contrast effects are best when light reflected from sign characters differs in intensity by as much as 100 percent from the background. This is exemplified by the use of black print on a white background.

Pictorial Signs

These are important where language barriers may operate. As much as possible, all pictorial symbols should be standardized. This has recently been accomplished for automobile dashboards, and is becoming common for highway signage.

Auditory Displays

The technology of auditory displays is rapidly developing through electronics and computer technology. Common auditory signals in the past have included beepers, buzzers, horns and sirens.

Auditory displays can be used effectively:

1) To reduce the workload on vision.
2) To provide an excellent attention-getting capability.
3) To incorporate immediate intelligence, as in voice signals.

However, unless carefully designed, auditory displays can introduce distracting clutter in an adverse noise field.

Recent developments in voice generation technology have been made in reading aids for the blind, speaking aids for the deaf, computer-aided instruction (CAI) for the general student, and computer interfacing techniques. Auditory techniques are also being used in computer input technology, but require further improvement so that the computers will respond accurately to different voice characteristics and accents.

In computer-generated speech, understanding is sometimes difficult because of the mechanical quality (Luce, et al., 1983; Slowiaczek and

Nusbaum, 1985). Tape recordings of natural speech are, in fact, more easily understood than is synthetic speech (Pisoni and Koem, 1982).[6]

Voice signals can yield immediate display intelligence as compared to simple alerting sounds such as bells or buzzers. Thus, the HO's coping and action responses can be markedly enhanced as important design considerations (Burgess, 1983).

Audio signals can be most appropriately used in design when:

1) The HO is mobile and visual messages are not readily perceived.

2) The HO has an excess of visual displays in the work environment.

3) Immediate attention is required.

4) Only short message signals are needed and can be discontinued after the initial advisory messages are acknowledged.

5) Light levels are too low for visual displays.

Summary

Human factors control-display design principles can be applied to virtually any product at any level or complexity. The first order of business for a designer is to determine what the HO must do and what information is needed to do it. Then and only then can control-display requirements be established. Control design principles to be applied can be analyzed in terms of the types of control responses needed and the best control configuration that can be provided to perform these responses. An information requirements analysis can then be completed to determine the kind and depth of feedback information the HO must have for managing the control task.

Displays can be aural or visual; they can be discrete or can provide continuous feedback information to the HO. Efficient labeling must be provided the HO to assure a clear understanding of what a control does or of what information a display provides.

---

6. Synthetic speech is generated through phoneme-based words, duplicating the human voice as produced by lips, tongue, teeth and palate. The sounds include (1) plositives produced by sudden air stoppage, e.g., "plop," (2) frictives, as when air-passage slits are narrowed, e.g., in the word "their," (3) laterals when air passes around a closed centerline, e.g., the word "lit," and (4) vowel sounds produced by unrestricted air passage (Miller, 1963).

6
Layout Design Principles

In the design process, after control-display requirements have been established and human factors design principles have been applied, a suitable and efficient layout configuration must be developed. This is so regardless of the level of product complexity, i.e., whether a simple tool, a simplex or complex machine, or a system is involved.

The designer must begin by asking certain fundamental design questions such as: What is the population of users? Are male or female users or both to be accommodated? What anthropometric design strategy should be employed from a cost standpoint in compromising the scope of user efficiency?, etc.

Layout Design Philosophy

A designer can approach the layout design from several different perspectives. In practice, the anthropometric design approaches used are usually the following:

1) A designer uses his or her own body dimensions for the design on a kind of "this feels about right" basis. At best, this approach is only randomly effective, and, at worst, is grossly inadequate for accommodating most of the user population.

2) A designer uses the average dimensions of a user population. This is a most common practice in the design of such items as kitchen cabinets, dressers, office desks, appliances, and other such products that

would just be too costly to produce in different sizes. The average accommodates a large percentage of the population, but becomes increasingly inconvenient for users who deviate more widely from an average size.

3) The extreme dimensions of user body sizes are used for the design. The designer must decide how extreme the dimensions should be, i.e., the 5th or 1st percentile, the 95th or 99th percentile, etc. In this approach, large sizing is provided for clearances such as in doorways, desk dimensions, and work areas. The smallest sizes are used for reach and viewing accommodations. When the extremes have been accommodated, all average or intermediate body sizes are also included. The dimensions are fixed in this approach, so the design is not necessarily too expensive to undertake. (Certain dimensions, such as work surface heights, of course, are not accommodative for the range of body sizes without being adjustable.)

4) The design of an adjustment range is an approach that accommodates a greater number of user-population body sizes.[1] Thus, straps, bands, etc. are used to adjust personal equipment for different head and body dimensions; adjustable seats and platform accommodations are other examples of accommodating a range of body sizes. An alternative course is to provide fixed sizes for the middle, upper and lower dimensions of body size. Attempting to accommodate a complete range of body sizes can of course be costly to implement in the fabrication or production process as compared to one fixed size of a product.

The Use of Anthropometric and Biomechanical Data in Tool Design

Body dimensions and capabilities data should be applied in the design of such products as hammers, shovels, push-pull carts, canes, crutches, etc. to assure a wider accommodation for a user population. Figure 6.1 presents a number of anthropometric dimensions that can be useful in designing handtools and hand-controlled machines. Hand dimensions, for example, can be applied in sizing tools for secure grasping. Tools designed for use by females as well as males should accommodate the

---

1. A wide range of body sizes for accommodation may be most important in the design of passive and active safety restraints, e.g., seat belts, air cushions, etc.

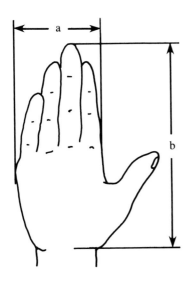

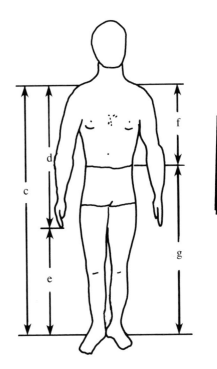

PERCENTILES (IN INCHES*)

	5th		50th		95th	
	Male	**Female**	**Male**	**Female**	**Male**	**Female**
Hand breadth	3.2	2.8	3.5	3.0	3.9	3.3
Hand length	6.9	6.5	7.5	6.9	8.2	7.5
Shoulder height	52.6	48.8	56.5	52.5	60.7	56.6
Shoulder-to-fingertip	28.2	22.9	30.5	25.2	32.8	27.4
Fingertip-to-floor	24.4	25.9	26.1	25.0	27.9	29.1
Back height						
(waist to shoulder)	14.2	12.0	14.6	12.6	15.3	13.1
Waist height	38.4	36.7	41.9	39.8	45.4	43.4

* Divide by .3937 to obtain centimeters.

Figure 6.1. Anthropometric dimensions useful in
tool-design applications.

5th percentile female hand length of 6.5 inches and breadth of 2.8 inches. In accommodating the female hand, the following tool dimensions can be prescribed:

- Maximum diameter for holding—3 inches
- Diameter for simple manipulation—2 inches
- Diameter for thumb-finger interlock—1 to 1.5 inches
- Diameter for thumb-finger precision control—0.25 inches

The hand breadth dimension used in hand-grasp design should consider the larger dimensions as well. The 95th percentile male hand breadth of 3.9 inches, for example, might be used.

These dimensions would accommodate both the small hand length of the female and the large hand breadth of the male.

Handle Contours

The shape and size of handles used in tool design should largely be a function of the force required and the direction in which it is to be applied. Cochran and Riley (1986) tested a number of handle configurations for different applications by male and female subjects. The handle contours used were (1) circulars, (2) circular with one flat side, (3) circular with two flat sides, (4) triangular, (5) square, and (6) rectangular.

Diameters varied from 0.8 to 1.6 inches. All handles were about 8½-inches long. Required arm postures tested were in the standing and sitting positions for force application as follow:

1) Arm extended down at the side while seated, swinging forward and backward.

2) Upper-lower arm angle at about 90° while seated with force exerted forward and backward.

3) Arm extended down at the side while standing, with torque force exerted clockwise and counterclockwise.

The investigators concluded that the design of handle contours should be based on the particular type of force required and the direction in which it is to be applied. The triangular shape, for example, was found to be most effective in the seated position with the upper-lower arm at 90° pushing forward. The flat side of the triangle produced a good surface contact at the area of the palm with a comfortable surface around which to wrap the fingers.

The design of tool handle contours and dimensions can be best based on the designer's own understanding of the force requirements and the directional aspects of the task at hand.

Other Body Dimensions in Tool Design

Other body dimensions should be considered in such products as push-pull carts and shovels. The "fingertip-to-floor" smallest dimension, for example, can be used in designing a cart handle length in the pull-push angle. The hypotenuse or handle length should thus be about 33 inches, i.e., 29 inches to fingertip, plus about 4 inches for hand grasping.

Awkward and fatiguing body postures, of course, should be avoided in tool design. Handle lengths for shovels, for example, should be sufficient to accommodate an upper-lower arm length of 80° to 120° in the working position. This places the handle of the angled shovel at about waist-level height for the 5th percentile female.

As previously noted, other undesirable fatiguing muscle tensions can be reduced by locating bends in the tools themselves rather than in the body-posture requirement, i.e., in pliers, wrenches, screwdrivers, hammers, etc.

Biomechanical force-exertion capabilities should also be considered in layout. For example, the right arm force in the 5th percentile male push-pull exertion is about 50 pounds when the arm is at waist level. When the arm is down, the strength is reduced by about half (refer to Table 3.1). To allow for a small female's force exertion capability, handles requiring a push-pull action should be limited to around 20 or 30 pounds, and should be located 36.7 inches high (the 5th percentile female's waist height).

Workstation Dimensions

In the design of work areas, control consoles, vehicle control stations, etc., anthropometric dimensions must also be taken into account. In many cases, the "extremes" of anthropometric design strategy can be used, with the smallest body dimensions employed to accommodate for reach-and-see functions and the largest for clearance accommodations. At the outset of design, it must be determined if both sexes are to be in-

83

volved in the operation. If so, the small female dimensions should be used, i.e., the 5th percentile. In some cases, the "extremes" approach should be combined with the "range-adjustment" approach or anthropometric strategy to assure maximum accommodation.

Figure 6.2 illustrates the most common workstation positions—sitting, sitting-standing, and standing. Table 6.1 lists some common workstation dimensions in these three positions for both male and female workers. The viewing angles indicated are off the horizontal plane. All measurements in sitting positions should be made off the seat reference point (SRP) illustrated, and should include all angular, horizontal and vertical dimensions pertinent to the control or work operations.

Table 6.1. Some common workstation dimensions (in inches*).

Position	Male	Female
Sitting:		
Work surface height (H):		
General	29	27
Precision work	40	36
Assembly work	36	33
Light work	30	28
Typing	28	26
Reach distance (R)	27	25
Viewing angle	60°	65°
Sitting-Standing:		
Work surface height (H):	37	34
Viewing angle	45°	50°
Standing:		
Work surface height (H):		
Precision work	45	42
Light work	41	36
Heavy work	36	34
Overhead reach (OR)	78	73
Viewing angle	30°	35°
Manual area (M)	45 to 57	42 to 54
Visual area (V)	52 to 65	49 to 62

* Divide by .3937 to obtain centimeters.

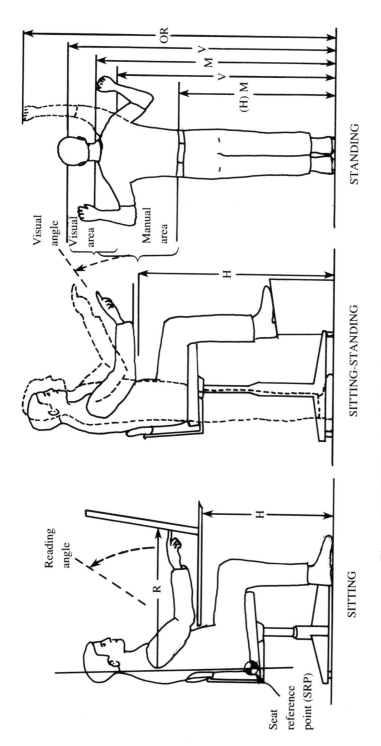

Figure 6.2. Workstation design dimensions (refer to Table 6.1.).

85

Principles of Layout Design

In preparing a layout design plan, basic anthropometric considerations must first be made. General principles in proceeding with the layout design should include the following:

1) Overall work space must accommodate the HO to include personal clothing and protective equipment.

2) Work space requirements should be determined from a study of body sizes of the using population. Gross dimensions should allow for the largest HO in the using population when encumbered with environmental protection clothing and equipment. The area should provide clearances at the knees, shoulders, hips and head while the largest HO executes full excursion movements required in task performance.

3) The work space should accommodate the see-and-reach envelope of the smallest-size operator as well as the largest. Within vehicle compartments, the small HO should be able to reach controls from within the constraints of a locked safety harness if so required.

4) Displays and controls should be prioritized on the basis of ranked importance, and based on linkage frequencies in turning from one control/display unit to another in normal and emergency operational sequences.

5) Upper and lower leg angles should be studied for sufficient leg and "kick" room. HOs should be able to stretch out their legs.

6) Upper and lower arm angles should be studied when operating controls. (Angles less than 90°, for example, can result in fatigue and cramps.)

7) Workstation distractions should be avoided. Open work areas involving such disturbances as personnel talking, high noise levels, poor lighting, and so on should be avoided.

8) Ample room should be provided for maps and drawings to be spread out if necessary. Poor reading postures can result in fatigue and poor comprehension. Maps and drawings may best be configured into indexed books or projector systems to avert this problem.

Completing analyses of operating situations can provide significant layout design data in providing for operations at critical times in a job sequence. A precisely integrated control-display arrangement, for example, may be necessary to assure that all operations can be carried out within the time limitations. In fact, all such time-limited sequences should be studied to assure that the layout is suitable for operations within the time constraints.

Designing for Force Exertion

A low seat position should be provided (approximately 7 inches high) when considerable leg force must be exerted from the seated position. Locate controls so that combined sets of muscle groups can be used to provide back and shoulder structural supports.

Designing for Adequate Control Location

The compartment should be provided with convenient control movement and easy see-and-reach activities. Structural frame members of the compartment should not obstruct control movements. Spaces between controls should be sufficient to prevent the HO from inadvertently hitting and actuating adjacent controls. The HO's body members in control movement should be free to move over the complete range of control actuation. Controls should be designed for "no-look" or blind operation, i.e., they should be tactually coded and located at front and center at about elbow level if blind reaching is required. Personal clothing and equipment should not interfere with control actuation.

Primary and Secondary Visual Areas

Primary visual areas make up the center points of the HO's direction of attention. These, for example, may be a straight-ahead outside visual scene or a CRT layout along the line of sight. These establish the central visual reference points.

Secondary visual areas are controls and displays adjacent to the primary visual area. In some operations, secondary displays are also being placed in the primary visual areas in so-called "head-up" displays. In aircraft, instrument display readouts are, in these cases, projected on the windscreen.[2]

---

2. Head-up displays have also been introduced into automobile stations. The numbers of a digital speedometer, for example, may appear to float in the air about half way down the length of the hood. The image presumably does not interfere with the driver's view of the road while permitting him to check his speed without looking down.

In fighter aircraft, information is displayed within the pilot's field of view so that he does not even have to adjust the focus of his eyes during combat.

The advantages of head-ups displays in automobiles, however, have not as yet been demonstrated. Some Japanese manufacturers have mounted liquid crystal displays on their windshields (American designers hold that head-up displays must, rather, involve a "virtual image.") Speedometer readings are projected from within the instrument panel onto a small holographic mirror mounted on the lower part of the windshield. The mirror is treated to reflect light in only a narrow band. The mirror is thus transparent to

Primary visual areas should *never* require major head or eye movement away from the normal line of sight. The latter is taken to be 10° down from the straight-ahead position as a visual comfort zone. All critical and emergency secondary displays should be within the 30° cone of the normal line of sight.

The viewing distance to a console should minimize visual fatigue at around 28 inches. Displays should never be less than 10 to 12 inches away even for short-duration reading.[3]

Secondary displays can be located where visual parallax is not an important consideration.[4]

Biomechanical Principles in Layout Design

Here are some biomechanical principles in control layout and design:

1) Work should be assigned to body members according to their strength and ability.

2) Control layout should *never* require that a single body member do all the work.

3) A minimum of body mass should be used whenever possible. For example, finger-actuated controls should be used since they have a lower mass than the hands, the hands have a lower mass than the arms, etc.

4) Layout should provide for the work to be balanced for symmetrical and opposite-hand action about the vertical body.

5) Crossing hands in a task should never be necessary.

6) Smooth, continuous ballistic motions should be possible, with a minimum of limb acceleration to perform the tasks. This results in less effort. For example, sharp, angular motions should never be required around objects to perform a task.

---

all other light. The projected image appears to be in back of the actual surface upon which it is projected.

Such head-up displays hold much promise, but must be carefully studied from a human factors point of view for effective application. Stan Roscoe (1986) cautioned designers on the use of collimated or straight-ahead display elements used in head-up displays (HUDs). When the eyes are focused for a distant visual display upon which HUD elements are projected, in order to see the HUD elements intermittent focusing is required. This can result in loss of distant visual acuity and spatial orientation, which can often be critical in vehicular control tasks.

3. Visual fatigue occurs when such short-distance viewing is required.

4. In reading displays off the centerline, the head and eyes can be shifted with a fixed trunk over a range of 95° left and right, 65° up, and 75° down.

Postures for Layout

The HO's required postures of sitting, standing, or a mix of these will determine how the control and displays are to be arranged (see Figure 6.2).

Seated Posture

A sitting posture in workstation layout is best when the HO must have a high degree of stability and equilibrium during prolonged work periods, and/or when the feet must be used for control.

Standing Posture

A stand-up posture is required when the HO must be mobile to monitor controls and displays at different consoles, but is not required to execute precision control. Only simple foot-control action should be required in this posture. Standing over long periods of time is fatiguing and is not recommended.

Standing-Sitting Posture

This combined arrangement is best when the HO requires stability for precision-control operations while seated, but is also required to be mobile and to stand while monitoring large control panels. In such cases, high chairs or stools should be provided with footrests.

Environment

Environmental protection in the workplace must be provided within both natural and machine-generated conditions. The protection must be acceptable from both a health and safety standpoint as well as from a performance point of view.

Noise

Ear protectors and communication adjuncts are needed in high-noise areas. At 125 to 8,000 Hz frequencies, covering the ears with the hands or cotton inserts can attenuate the noise level by 20 to 40 decibels. Circumaural protection, such as helmets and earmuffs, can attenuate noise up to 48 decibels.

Vibration

Here, comfort and performance efficiency can be affected. Absorbents or shock-absorbing mountings should be used to reduce the effect of excessive vibrations.

Radiation

Protection may be required from X-ray, beta, and gamma ray radiation. When such levels present a risk, they should be frequently or continuously monitored. Protective clothing and head gear are necessary when working under conditions of radiation.

Lighting

Task data are needed to determine the visual detail involved and the light levels required. In some cases, levels as high as 1,000 foot-candles may be required (Faulkner and Murphy, 1973).

Atmospheric Conditions

Ambient pressures vary with altitude and under water. In low pressure, personnel require artificial pressure to control internal body pressures from "ballooning" or expanding in the body tissues. In high undersea pressures, special compression and decompression procedures must be used.

Temperature and Humidity

Optimal temperatures for a work environment are a function of the task. The relative humidity should be below 60 percent for proper body cooling. Gloves must be worn in temperatures below 55° F. for good circulation and finger dexterity.

Air Pollution

Careful measurements are needed for machine exhaust products and possible toxic effects. The HO may require tanked air and breathing equipment when exposed to dust and toxic gases.

Noxious Liquids

Protective measures are needed around hazardous liquids. Protective clothing and goggles should be worn. An easily accessible shower and eyewash facilities should be made available when the HO can possibly be sprayed with caustic or other harmful solutions.

Motion

This may be rotational or linear in nature. Accelerations can cause disorientation and motion sickness. Special supports may be needed at the workstation under acceleration to prevent interference with the HO's control movements.

Summary

After the controls and displays of a product or piece of a equipment have been designed, they must be arranged to optimally facilitate an HO's performance. Various anthropometric design practices may involve the designer using his or her own body dimensions to arrange a layout, the use of average user body sizes, the use of extreme user body sizes, and the use of a full range of user body sizes in designing adjustability into the product layout. Layout design principles must be incorporated into product design at any level of complexity; this is essential if the HO is to perform efficiently with the product. The designer must also consider environmental protection requirements as part of the layout design.

DESIGN APPLICATION

This form is meant to help the student bring together the principles of human factors design as outlined in the previous chapters.

HUMAN FACTORS DESIGN APPLICATIONS

Controls	Location on Unit	Human Factors Principles Applied

Displays	Location on Unit	Human Factors Principles Applied

DESCRIBE LAYOUT AND OPERATING POSTURE REQUIRED: _____

_____

_____

_____

DESCRIBE ENVIRONMENTAL CONDITIONS AND PROTECTION REQUIRED: __

_____

_____

_____

DESCRIBE USER TESTS COMPLETED (mockups, simulators, prototype equipment, etc.):

Control-Display Testing Completed	Recommendations
Environmental Testing Completed	
Layout Testing Completed	

IV
Product Improvement

7

Human Factors Applications in Product Reviews

The basic principles for control and display design and layout in simplex and complex products were discussed in Chapters 5 and 6. Now, several simplex products will be examined, where pertinent, in the application of these principles. The products have been on the market and in use for some time. Some products may have been poorly designed from a human factors point of view and, as a consequence, may even be losing a share of the market for this reason. A human factors review procedure thus becomes of very real concern when considering product improvements.

Task Versus Activity Analysis

A distinction should be made between the type of analysis made when the product configuration is already firm and operational, and the analysis required when the product is only under development with the controls and displays yet to be specified. When the product is operational, a "task analysis" can be completed; the precise control action required can be described for a specific control, and the display readings provided in the configuration can be outlined in detail on a task-by-task basis. A task analysis procedure, strictly speaking, is largely limited to product-review situations; or, for the purpose for which it was originally formulated by Robert Miller in the 1950s, the procedure can provide the kind of detailed information necessary for the design of training programs and simulators.

An activity analysis, on the other hand, is required when a product configuration does not currently exist. The analysis must, therefore, be carried out to establish control-display and layout requirements. The activity-analysis type of procedure will be discussed in Chapter 8.

Task Analysis

The task analysis procedure is quite extensive and detailed as originally formulated by Miller (1953). It essentially entails a sequential description of time-shared tracking (continuous) and procedural (discrete) tasks.

The following are steps to be completed in the analysis:

A. Describe the tracking tasks.

   1) Describe controls and displays.
   2) What are the critical parameters?
   3) What are the time-related values involved?
   4) What are the critical-delay functions?
   5) Describe the critical indications.
   6) Describe the corrective actions required.
   7) What are the likely typical errors?
   8) What are the situational stresses?
   9) Describe the typical decisions to be made.
   10) What is the significant feedback to the HO?

B. Describe the procedural tasks.

Since automobile driving is such a commonly understood machine operation, this will be used, as Miller used it, to illustrate the task analysis procedure as follows:

A. Describe the tracking tasks: Maintain direction and velocity control within roadway traffic regulations.

   1) Describe controls and displays.
     Controls: Steering wheel and foot brake linkages connected to ground wheels; foot pedal connected to carburetor.
     Primary displays: Yellow and white lines left of center; vehicles oncoming at left, following at rear, leading at front; parked to the right. Visual scene motion varies with velocity.

Secondary displays: Speedometer, gas gauge, and operating-condition indicator lights.

2) What are the critical parameters? Keeping to the right side of the road, following regulatory signs, and avoiding vehicle collision.

3) What are the time-related values? Speed, braking rate, and fuel-consumption rate.

4) What are critical-delay functions? Vehicle acceleration and brake pedal deceleration lags.

5) Describe critical indications. Motor noise, clearances for other vehicles, rate of scene change, speedometer pointer below legal speed limit, road lines to the left.

6) Describe corrective actions required. Steering wheel regulation for proper course correction; accelerator pedal and brake pedal regulation for continuous velocity correction.

7) What are typical errors? Failure to judge for proper velocity and to anticipate precise braking response required; moving too far in or out with respect to tracking course in risking collision.

8) What are the situational stresses? Traction impaired by slick road surfaces, vision impaired by fog and darkness; multiple decisions with high degrees of uncertainty resulting in vehicle collisions.

9) Describe typical decisions. When to proceed through stop signs and to enter into traffic patterns; what the best traffic patterns are on one-way streets; when a door comes loose whether to stop and close it; when to speed up or slow down.

10) What is the significant feedback? Changes in the visual scene; the springy "feel" of the steering wheel and foot pedals; speedometer pointer rising and falling with accelerator foot pedal action.

B. Describe the procedural tasks. Perform all turn-on and shut-down subtasks in vehicle operation as required. Example:

1) Subtask: Turn on headlights.

    a. Control: 3-position, stalk-mounted rotary "On-Off, Park" detent control switch.

    b. Action: Rotate control knob 45° through "Park" detent position to "On."

    c. Display: Forward-looking light beam illuminates scene; "Hi" beam amber indicator light glows.

d. Objective response: Road objects become clearly visible for safe steering response.

e. Possible malfunction: Light circuit "shorts out"; bulb burns out; switch detent position slips.

Product Review Procedures

In product-review cases, particularly where simplex products are involved, the task analysis procedure may be somewhat simplified from that of the above. However, review procedure should include at least the following:

A. A description of the product.

B. A task analysis of it.

C. A description of the environmental conditions of its operation.

D. A product assessment from a human factors point of view.

E. A listing of design-improvement recommendations with their rationales.

Following are simplex machine products that have been subjected to the product review procedures described above in order to illustrate their applicability. These have been selected neither because they are especially good or especially poor from a human factors point of view; rather, they have been included for review simply to show how just about any product can be subjected to such a product-review procedure.

K-Mart electronic LED alarm clock, Model KMC-521

A. Product description: This is a solid-state, four-digit LED timing display with a beep alarm. The unit is illustrated in Figure 7.1. (Pushbuttons are knurled, which is not evident from the illustration).

B. Task analysis:

a. Setting time:

Depress "Fast" pushbutton while holding down "Time" pushbutton for fast change of digit reading. Depress "Slow" pushbutton for slow change of reading.

Note: After a power interruption, the digits will flash on and off until time is reset. "PM" or "AM" settings must be observed to assure 24-hour memory.

b. Setting alarm:

1) Place "Alarm" switch in "Off" position.

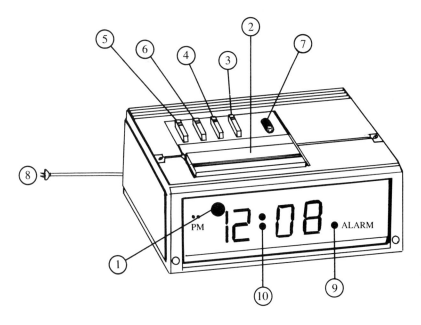

1. PM indicator
2. Snooze button
3. Time set button
4. Alarm set button
5. Fast set button
6. Slow set button
7. Alarm on/off switch
8. AC power cord
9. Alarm on indicator
10. Clock display

Figure 7.1. K-Mart electronic LED alarm clock, Model KMC-521.

2) Depress "Fast" pushbutton while holding down "Alarm" pushbutton for fast change of digital reading. Depress "Slow" pushbutton for slow change of digital reading.

3) Place "Alarm" switch in "On" position.

c. Turning off alarm when sounding: Place "Alarm" switch in "Off" position.

Note: Immediately repositioning switch to "On" position will reset for next day, since the unit has a 24-hour memory.

The "Snooze" pushbutton is used for nine-minute intervals of sleep, continuing for up to an additional hour.

C. Environmental conditions: The timer is used in a darkened bedroom environment. Setting operations are made while the HO is alert, but alarm shutoff sequences are executed in a "sleepy" or drowsy condition of the user.

D. <u>Critique</u>:
   1) Labeling: All labels are below or to the right of controls and displays except "Snooze" and "PM" which are inconsistently located.
   2) Functional grouping: Alarm functions are mixed in with those of time. These should be rearranged.
   3) Indicators: The "Alarm" indicator light seems unnecessary if an adequate switch position reading is provided. Red lights should be reserved for emergencies.
   4) Alarm shutoff: When the HO is drowsy, the "Alarm" slideswitch does not seem large enough for quick positioning to the "Off" position, thus taking longer to be shut off. Both hands may be required to operate the "Alarm" switch fore and aft since the lightweight unit moves with the switch action unless held in place. The switch position also does not provide for good status reading.

E. <u>Recommendations</u>:
   1) Labeling: Locate "Snooze" and "PM" labels below lights. Reason: Consistency should make for easier reading and prevent confusion.
   2) Functional grouping: Group "Alarm" setting pushbutton with alarm "Snooze" bar and "On-Off" switch. Reason: Functional grouping of controls and displays improves readability and clarity.
   3) Indicators: Eliminate the "Alarm"-on indicator light. Reason: Switch position provides the same indication. Simplification, where possible, is desirable for improving a task.
   4) Alarm cutoff: Use a rocker switch instead of the "On-Off" alarm slideswitch. Reason: The up-down rocker motion of the switch is less likely to move the whole unit when the switch is actuated. Also, a rockerswitch provides a better position status reading than does a slideswitch.

<u>K-Mart Execucard solar-powered calculator, Model KMC-TM3</u> (Figure 7.2)

A. <u>Description</u>: The unit is designed as a credit card-sized calculator powered by an outside light source through a solar cell.

Figure 7.2. K-Mart Execucard solar-powered calculator, Model KMC-521.

B. Task analysis:
1) Expose solar unit to light source.
2) Clear the unit by pressing "ON/C" and "AC" areas.
3) Press numbers for calculations. Position decimal point (clear erroneous entries by pressing "ON/C" once; press twice for total clearance); press symbols as needed for addition, subtraction, division, multiplication, and to compute square root, reciprocals, and percentages.
4) Store solutions in memory by pressing "M-" or "M + ".
5) Call up memory by pressing "M_C^R".
6) Clear memory by pressing it again.

C. Environmental conditions: Light sufficient to read by is necessary to operate the solar cells. The unit may be operated in temperatures above freezing.

D. Critique:
1) Punching: The characters are merely pressed and must have immediate visual readout for feedback. Different pressure areas seem to operate, i.e., some numbers are difficult to operate.
2) Layout sequence: Left to right, top to bottom operational sequences do not seem to have been adequately studied for a more efficient layout.
3) Number consistency: Number layout is not consistent with the layout of telephones, resulting in a poor transfer of punching habits.
4) Key duplication: The all-clear "AC" pressure actuator function duplicates that of the "ON/C" actuator.

E. Recommendations:

1) Displacement pushbuttons: Use controls that displace in actuation rather than the pressure actuators. Reason: Positive feel of control will then be provided.

2) Combining controls: Combine the "AC" key with the "ON/C" key. Reason: Key duplication in functional tasks seems unnecessary.

3) Layout of controls: Change the arrangement of keys to improve sequences in the more complex operations, proceeding from left to right and top to bottom. Reason: This arrangement is in keeping with convention. Operating time can also be reduced by such an arrangement.

4) Eliminate unnecessary symbols: Trade names should be deleted from the face. Reason: Unnecessary symbols distract the HO. Trade names are better placed on the back for this particular model.

5) Control space: Controls should be separated for more space. Reason: Wearing gloves when operating controls becomes more feasible.

6) Key layout: Sequence of keys should conform to that of the standard telephone keyboard. Reason: Keypunch habits would be facilitated.

Radio Shack AM-FM Shortwave portable transistor radio, Model 3414 (Figure 7.3)

A. Description: The product is a solid-state, 14-transistor, AM-FM and shortwave radio, with rotary tuning, tone and on-off/volume controls. Slideswitches include "AM-FM-SW" and "AFC-OFF". A fold-in antenna is also provided.

B. Task analysis:

1) Adjust antenna.
2) Select mode of operation: position "AM-FM-SW" slideswitch to desired mode.
3) Turn on power.
4) Adjust volume and tune in station.
5) Adjust tone control.
6) Turn on AFC for FM frequency modulation.

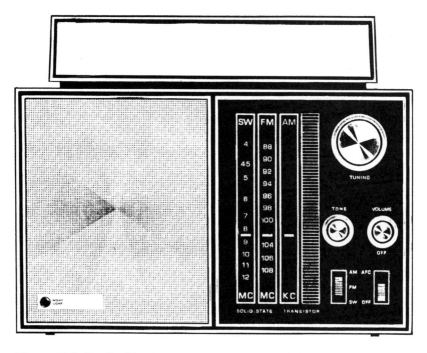

Figure 7.3. Radio Shack AM-FM-Shortwave portable transistor radio, Model 3414.

C. <u>Description of environment</u>: The radio can be used indoors or out-doors, in light or darkness, and in temperatures where gloves are worn.

D. <u>Critique</u>:

1) Grasp of controls: Knob depth is insufficient for easy grasp, par-ticularly if gloves are worn.

2) Control layout: Controls located to the right of tuning dials would be blocked if left hand is used.

3) Labels: These are inconsistently located above and below and to the left and right of controls.

4) Number of dials: Two megacycle tuning dials seem unnecessary. These could likely be combined into one

5) Control status: Position status of volume and tone rotary controls is not evident. Also, the slide switches do not clearly indicate po-sition status.

6) Unnecessary markings: Nomenclature of "solid state/14 transistor" is more appropriate for inclusion in a manual.

E. Recommendations:

1) Locate controls below the displays. Reason: Either left- or right-hand operation will not block visual access to the display.

2) Lay out controls for sequential operation from left to right, starting with the mode selector switch, then the on-off/volume adjustment and tuning selector. Reason: Left-to-right sequence is in keeping with normal user expectancies.

3) Locate all labels consistently. Reason: This avoids confusion, and makes for easier reading.

4) Employ pointer-selector knobs as adjustment controls for the mode selector and on-off volume controls. Reason: Position status can then be efficiently read.

5) Use ganged controls for tuning. Reason: The outer control knob can be used for slewing, and the inner for fine tuning.

6) Increase control depth. Reason: This provides for easier grasping with and without gloves on.

Hotpoint automatic clothes washer, Model WLW 2500 B

A. Description: The washer provides an automatic cycling of washing, rinsing and wash-spin for garments of cotton and linen, permanent press garments, polyester knits, and synthetic and blended fabrics. Controls include a temperature selector with a detented three-position knob, a detented three-position water-level knob, a two-position rotary selector knob for setting the speed and a rotary cycle-selector knob. The configuration is illustrated in Figure 7.4.

B. Task analysis:

1) Determine size of load.

2) Set water-level knob for less than half a load ("Low"), between one-half and two-thirds ("Medium"), and over two-thirds ("Full").

3) Set temperature according to fabric(s) and how extensively soiled the material is.

4) Set spin speed for washing and spin dry.

5) Set washing cycle by fabric and extent of soiling.

C. Environment: The hands may be wet and perhaps soapy due to a liquid cleaner.

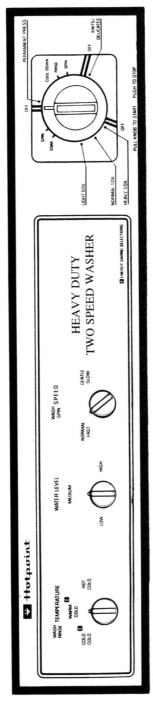

Figure 7.4. Hotpoint automatic clothes washer, Model WLW 2500B.

D. Critique:

    1) A more logical layout sequence might be the load (water-level) setting first, then the temperature, and then the spin speed, i.e., the water level should come first.

    2) The use of a selector rotary knob for only two spin speed positions is inappropriate.

    3) The spin speed selector that increases the speed counterclockwise is contrary to convention which increases in the clockwise direction.

    4) The cycle selector uses three separate timers. These could be integrated into one or possibly two timers.

    5) The cycle-selector labeling seems excessive.

    6) The cycle knob should indicate that it must be pushed to be set.

    7) Knob surfaces are smooth chromium which could be slippery when the hands are wet.

    8) Caution notes should be included to advise users on how to prevent damage to garments or to the machine.

E. Recommendations:

    1) Use a "Load" selector knob rather than a "Water-Level" control. Reason: The user's selection decision is based on the size of the load, which then has to be translated into the water level. A "Load" nomenclature, with "Small," "Medium," and "Large" positions would be directly related to the size of the load.

    2) Lay out controls in a logical sequence, proceeding with "Load," "Temperature," "Spin," and, finally, "Cycle." Reason: This should conform with expectancies and enhance ease of reading.

    3) Use pointer knobs for the load and temperature settings. Reason: These would provide more positive position status readings.

    4) Use a rocker switch for the spin setting. Reason: A rocker switch would lend itself more readily to a two-position setting than does a rotary selector switch.

    5) Arrange the "Cycle" selector positions so that the more frequently-used cycle positions come first. Reason: The user's task would be simplified with setting time reduced.

    6) Simplify the labeling. Reason: Confusion often results from an excess of labels.

    7) Provide caution and warning notes where damage to garments or the machine might result, e.g., not to use a hot setting for knit or

polyester garments, not to turn the cycle rotary switch counter-clockwise or else damage will result, etc.

Additive System Units

Products are often added to an operational system that already has on-going task requirements, some of which may be quite demanding. An automobile, for example, is a system with ongoing driving tasks. Products such as cellular telephones, navigational computer maps, or police radar detectors thus require that their operational task requirements be superimposed onto and integrated with the ongoing driving tasks to insure safe and efficient operation.

A police radar detector has been selected to illustrate the kinds of design considerations that should be made when integrating new products into ongoing systems. Radar detector units are, of course, only quasi-legal; in some states, they are even prohibited or are regulated by local and state laws. However, apart from the legal and ethical implications in using such units, their system value has also yet to be adequately assessed. For example, does the unit merely give a license to speed, or does it heighten driver alertness and caution? System questions can perhaps be best answered by addressing specific evaluation criteria such as accident rates. In any event, the unit is included here as an analytical example since it presents a good design illustration of an additive unit to be integrated with the automobile system.

Bel Tronics Microeye radar detector, Model XKR-VII

A. Description: The detector is an electronic X and K radar band receiver, designed to alert a driver to police radar tracking. The unit is operated by being plugged into the vehicle's cigarette lighter socket for DC power. It includes controls and displays as illustrated in Figure 7.5. The six red light-emitting diodes (LEDs) light up as the signal strength from police radar increases, as the signal grows stronger, more lights glow together with the increased rate of a clicking sound. A yellow LED light comes on when the unit's power is initially turned on, and flashes when the LR and RSD modes are actuated. The flashing occurs differentially to distinguish the modes.

The reception-at-short-distance ("RSD") "Lo" position reduces

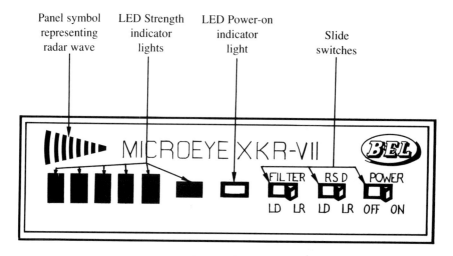

Figure 7.5. Bel Tronics Microeye radar detector, Model XKR-VII
(approximately to scale).

sensitivity to X-band reception in cities. (The latter is also activated
by microwave ovens and garage door actuators). The "Filter" mode
laterally operated toggle switch also filters out X-band signals of
low-energy levels when in the "Lo" position. The long-range ("LR")
position for both switches increases reception sensitivity to X-band
and K-band radars. The "Power" switch on the unit energizes it.
B. Task analysis:
   1) Plug in the unit.
   2) Move the "Power" switch to the "On" position.
   3) Select the "Filter" mode. (Use the "LR" position when on an
      open road.)
   4) Select the "RSD" mode. (Use the "LR" position when on an
      open road.)
   5) Check the red LED lights for increasing signal strength.
   6) When an increasing X-band or K-band police radar signal
      strength is indicated, check the speedometer and adjust the speed
      if excessive.
C. Environment: Operation of the unit is under day and night conditions in the interior of the car. The unit is also subjected to vehicle
   accelerations and decelerations. Gloves may be worn in its operation
   during cold outside temperatures.

D. <u>Critique</u>:

   1) Operating sequence in switch layout is right to left rather than the conventional left-to-right layout.

   2) Visual indicator reading tasks imposed on the driver can be unnecessarily distracting and perhaps dangerous. (Audio signals alone would be preferable.)

   3) Yellow LED switch position light would be unnecessary if the position of the switch, per se, could be clearly read.

   4) Labeling and abbreviations used may not be clearly understood.

E. <u>Recommendations</u>:

   1) Use a two-position toggle switch for "Power." Reason: Position displacement provides a visual position indication and eliminates the need for LED indicator lights which can be distracting.

   2) Use a two-position "City-Hwy" toggle switch. Reason: Providing positive switch-position reading eliminates the need for the LED switch-position indicator light.

   3) Integrate the filter switch with the "City-Hwy" switch. Reason: The functions can be reasonably combined and task complexity for the driver is thus reduced.

   4) Use a beep signal with a test pushbutton and volume control for radar detection above a set threshold value. Reason: Audio signals are best as attention intruders and do not require monitoring, thus eliminating the distracting visual signals imposed on the driver.

   5) Move trade name labels to the back of the unit. Reason: Unnecessary markings on the operating face of the unit are distracting to the operator.

Redesign of the Common Toilet

A widely used product that could reasonably be redesigned is the common toilet or western commode. This is essentially a porcelain water closet with a flushing mechanism and an open seat used for urination and defecation. The toilet seat is contoured and standardized for the bare posterior of male and female adults. (Seat adaptors are used for small children). Standard toilet dimensions are illustrated in Figure 7.6

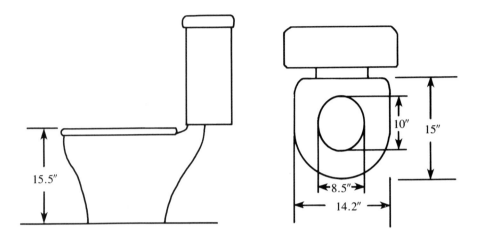

Figure 7.6. A standard toilet configuration.

How the Toilet Is Used

Male use of the toilet in urination is usually from a standing position, while female urination is done while sitting. In defecation, both assume a sitting position. In addition to elimination, a common use of the toilet is also for relaxation and such miscellaneous activities as reading while sitting in the defecation posture. Use procedures require putting the seat up or down for standing or sitting and, when finished, actuating a flush valve.

Environment

Use of the toilet may be under various conditions of lighting ranging from very bright to very dark. At night, users are usually in a sleepy condition and may not be dark adapted; they tend not to turn on the lights since they wish to retain the sleepy mood. Other conditions in the environment include the flushing noise and body gas odors.

Toilet Design Problems

The basic design of the toilet has changed little over the past several generations of users (*Human Factors Bulletin,* 1976; Kira, 1976; Mc-Clelland, 1982). Basic design problems have included:

1) Poor postures for defecation.
2) Male misdirection of urine while standing.

3) An improper fitting of the full range of adult body sizes in the standing or sitting position.

Defecation Posture

The posture required by the standard western-style toilet for defecation has been criticized as being unhealthy by promoting constipation (Kira, 1976; Hoffman, 1979). People in some parts of the world, for example, do not use toilets and simply squat in the open. In other cases, toilets are mounted on the floor, thus requiring that a squatting posture be used. A squatting position facilitates complete evacuation in exerting pressure at the abdomen. Such pressure is not applied when in a conventional sitting posture which is induced by the western-style toilet, and the resulting constipation can cause severe ailments. (The latter are usually attributed to stomach upset, however, rather than to poor elimination.) Constipation can occur from waste putrefaction, and may result in a number of symptoms ranging from mental depression to physical weakness, foul breath, a bloated stomach from gas accumulation, belching and flatulence, and, in extreme cases, appendicitis, dyspepsia, ulcers, colitis and colonic cancer.

A squatting posture is helpful in preventing constipation since the abdomen is compressed by the thighs, thus forcing the feces into the rectum. Conventional toilets do not accommodate a squatting posture. Bending forward induces some abdominal compression, but squatting greatly facilitates it. In fact, squatting appears to be a universal defecation posture except for those with western-style toilets.

Male Urine Misdirection

Toilet design also does not facilitate precise control of male urination in the standing posture. The area around the feet is sometimes insufficient for larger males, thus requiring a leaning posture over the bowl. This can result in the splashing or dripping of urine. The acid effects can result in floor damage and a malodorous condition.

Body Dimensions Accommodation

The toilet design also does not accommodate extremes and, in some cases, a majority of body sizes for sitting, relaxing and elimination. Several pertinent dimensions that should be considered include:

1) Buttock-popliteal length
   Maximum male—23.5 inches
   Minimum female—10.6 inches

2) Popliteal height, sitting
    Maximum male—21.3 inches
    Minimum female—13.2 inches
3) Hip breadth, sitting
    Maximum male—19.6 inches
    Minimum female—10.6 inches
4) Ankle height
    Maximum male—7.5 inches
    Minimum female—2.9 inches
5) Foot length
    Maximum male—12.3 inches
6) Foot breadth
    Maximum male—5.1 inches
7) Knee breadth
    Maximum male—4.7 inches

Recommended Design Features

An improved toilet design would include features discussed above. An improved posture for defecation, for example, could be better facilitated by the feet being raised or the bowel being lowered, thus permitting the user to assume a squatting position. In a squat, pressure would be applied to the abdomen that would better facilitate bowel elimination.

Male Urination

Accuracy in directing the urine flow from the standing position might be facilitated if more foot space were provided under the bowl. The male could thus stand closer without having to lean forward. Other approaches might include a urinary channel design on the underside of the seat. This would take advantage of the male's habit of raising the seat for urination.

Body Dimension Accommodation

Any proposed new toilet configuration should take into consideration the range of male-female body sizes, including the maximum buttock-popliteal length and hip-width dimensions—the largest male buttock-popliteal length dimension and the smallest female hip-breadth dimension. The seat height might best be at 17 inches to accommodate the largest number of users.

Design Tests

Any proposed changes in toilet design will basically require that extensive evaluation tests be made, using full-scale mockups. Careful physical observations and user assessment data should be employed with the mockups, which should be followed by the more extensive use of functional prototype models for continuing evaluation and design improvements.

The above examples of a product-improvement design approach are meant to demonstrate a procedure that can be used in the human factors assessment of varied and diversified products. The recommendations are not necessarily the final ones for design changes; they, too, must undergo careful scrutiny. Still, the recommendation format easily lends itself to further analysis. Requiring a rationale for each human factors recommendation provides a ready assessment of its validity; at the same time, if indicated, it can also be readily negated based solely on the merit of the rationale and on supporting evidence for the proposed changes.

Summary

A human factors product review procedure as described in this chapter can be applied to any product currently on the market. In cases where a product has been poorly designed from a human factors point of view, its very survival may become contingent upon significant design improvements.

Exercise 2—A Product-Improvement Review Exercise.

Perform a product-review procedure for a product of your own choosing, such as an appliance, a radio, a calculator, a machine shop tool, or any other such simplex product.

Your study should include the following steps:

1) Study the product and describe how it operates.
2) Provide an illustration of its current control-display configuration.
3) Outline the tasks in using the product. Include environmental conditions under which it is used, e.g., the light levels, whether used indoors or outdoors, with wet hands, etc.

4) Perform a critical analysis.

5) Illustrate, or prepare a cardboard mockup of, an improved config-uration. Provide a supporting rationale for each proposed change.

6) Organize and prepare a report that will include all the items out-lined above.

Study methods might include:

- Obtain the instruction manual for the product.
- Observe someone working with the product or perform the opera-tion(s) yourself.
- Query a user or someone thoroughly familiar with the product or ma-chine operation, e.g., someone who has owned the product for some time, a machine shop manager, etc.

V
New Products

8

Human Factors Applications in the Design of New Products

In the previous chapter, analytical procedures and examples were presented for a product-improvement review. Further analytical procedures will now be discussed that will permit a designer to proceed with a new product design. Project details can then be analyzed for human factors applications for a product that is only in the process of being developed. In the case of new-product design, the product may be of a simplex or complex nature. It may not necessarily involve wide-ranging implications wherein a total system approach is required, but, as a new product, as many human factors ramifications of its design should be considered as possible. Complex products with which designers might become involved include items such as diggers for deep mining, airport security stations for carry-on luggage inspection, self-service supermarkets which require customer-flow efficiency, or any of a variety of light, medium and heavy passenger, commercial and industrial vehicles. When such products are being initially considered in design, appropriate analytical procedures should be undertaken to assure that human engineering factors will be considered. A general analytical outline follows:

1) Describe the nature of the requirements and the purposes to be addressed by the product.

2) Complete a preliminary design study to determine essential background information about the design requirements and alternative methods of meeting the requirements; find out about any products that have been developed in the past to meet such needs.

3) Select the specific means to be developed for meeting the present need.

117

4) Identify specifications or the specific design requirements that the product must meet.

5) Prepare a list of activities and the sequences involved in operating the product.

6) Identify the kinds of controls and displays that will be needed by the HO to operate the product.

7) Show the control-display arrangements and environmental and safety provisions that are best suited to a population of HOs.

8) Outline the kinds of tests that will be carried out to assure the workability of the product before it is produced.

For illustrative purposes, let's take the example of requirements for ground-surface excavations and follow the analytical steps.

1) Requirements and Purposes

Ground-surface excavation equipment design is required for such projects as the surface mining of coal and various surface minerals such as phosphate. The purpose of its design is to provide the most reliable, safe, efficient and low-cost means of capturing these surface resources for commercial processing.

2) Preliminary Design Study

Prior to making pertinent design decisions about the excavation device to be deployed, background information must be acquired on excavation, its history, state-of-the-art equipment, and the human factors that must be considered.

Excavation involves stripping away soil and rock to expose the sought-after minerals. Surface excavation, as a technique, grew quite extensively in the 1960s in the surface mining of coal. Power shovels have frequently been used for excavation, as have draglines. Power shovels excavate in an outward motion, whereas draglines dig down and pull the excavation materials inward. Other earth-moving equipment used for excavation in the past included bulldozers and scrapers, many of which significantly increased in size during the 1970s. Vehicle equipment used to haul excavated materials also greatly increased in size during the 1960s and 1970s.

Power shovels have a somewhat better productivity capacity than draglines, but draglines have been more extensively used for economic reasons (Cone, et al., 1976). The smaller dragline excavation machines require less capital investment, and are often considered to be adaptable in different surface mining situations.

Human Factors Considerations.[1]

In excavation operations, numerous human factors problems must be addressed in design. HOs require a means of manipulating the digging mechanism, of communicating with personnel, training to operate the device, and protection from any adverse conditions in the operating environment.

Tracking Control

Basic to the design of any excavation machine is the HO's manipulation of the apparatus in vertical, lateral and longitudinal tracking control in the digging function. Ziegler (1968) found that the more complex the tracking task, the greater the workload for the HO, and the more tracking aids required. A free-moving control configuration, compared with a fixed control that reacts to simple pressure, can improve the tracking performance (Burke and Gibbs, 1965; Bryson, 1975). Sufficient tracking task practice is also found to enhance performance (Alderson and Whiting, 1974). Other studies have demonstrated that a control manipulator can be most efficient when the control is geometrically similar to the actual manipulator and incorporates force feedback (Kugath, 1973).

Communication

In the preliminary design study, functional transactions required among HOs must first be determined. These may involve coordinated actions by the rig operator and maintainers, land surveyors, truck haulers, etc. When these communication requirements are clarified, state-of-the-art communication equipment—radio microphones, wired apparatuses, etc.—can be investigated.

---

1. At the outset, human factors design research may involve only general considerations common to all design approaches. The research can then become progressively refined as specific methods of meeting the requirements are decided upon.

Personnel Skills

The most simple approach to personnel selection and training is preferable from a cost standpoint. If the necessary skills are readily available due to HOs using similar equipment, this might be a major design consideration. Thus, on-the-job training might be all that becomes necessary (Damos, et al., 1981). If more specific training and training equipment is required, this can become an additional tradeoff in making the design decision.

Layout Design and Environmental Protection

In the preliminary design study, and during the actual design phase, control-display design principles as discussed in Chapter 5 should be applied. The task load may require special layout considerations, such as the hand-travel distances required in control movements. The control layout will determine, for example, how frequently an HO is required to look away from the pickup points in a dragline tracking task (Mourant, et al., 1977). The reaching boundaries in control layout and provisions for seating comfort over prolonged work periods should also be considered (Bullock, 1974; Bendix, 1984).

The study should also include a detailed analysis of the inside and outside work environment. The outside environment in a digger operation, for example, includes temperature and humidity extremes, precipitation in the form of snow, sleet and rain, dust and dirt agitation, and operating in broken and rough terrain. The inside environment includes fumes from fuel combustion.

The general hazards of operating equipment such as mechanical diggers should also be carefully thought out. For example, malfunctioning hydraulic pumps or broken linkages in the shovel or scoop should be designed to be fail-safe, e.g., the shovel will not break loose or swing out of control when the apparatus fails, but, rather, locks into some safe position.

3) Means Selection

When the preliminary design study has been completed, sufficient information should then be available to make a means selection in the excavation process. Dragline bucket maneuverability, for example, has been found to have somewhat less precision than that of power shovels;

however, inherent excavation advantages of draglines have often favored the dragline approach.[2] Our selection in the excavation operations that we will deal with here will, therefore, be the dragline.

4) Design Specifications

Given a specific means approach to excavation, the general design attributes required can thus be specified. Dragline design specifications should include the following:

1) Efficient dragline manipulatability.
2) Optimal direct visual contact with bucket-control components.
3) A minimal and efficient control-display complex.
4) Fail-safe features incorporated throughout.
5) Provisions for operating in environments ranging from subfreezing temperatures to semitropical climates.
6) Minimal requirements for testing potential personnel out on the operations and for on-the-job training to acquire basic skills.

5) Activity Analysis

As stated previously, a distinction can be made between an "activity" analysis and a "task" analysis. In a task analysis, as discussed earlier, a firm control-display configuration has been established for the product which has been operational for some time. In an activity analysis, controls and displays are yet to be selected or designed since the product does not currently exist. Thus, in the latter case, the analysis must be based on projected activities which will then provide for a selection or design of appropriate controls and displays. In the dragline example, activities can be projected as follows:

The dragline scoops up the overburden and deposits it aside to expose the mineral material. It then scoops up the material which is either dumped into sumps to be pumped to the treatment plant or is deposited in a hauler to be transported to the processing plant. Specific activities include:

---

2. Refer to U.S. Bureau of Mines, *Miner Yearbook* for various years in the 1970s.

A. Start up and move vehicle into position for digging.
  Note: The dragline engine and rig service are checked out by main-
    tenance personnel. The dragline HO simply communicates
    with the maintenance people to verify readiness for operation.
    If engine and rig monitoring are otherwise to be required of
    the HO, these must be determined in the activities analysis.
  1) Move vehicle forward.
  2) Control left-right movement.
  3) Brake vehicle movement.
B. Lock vehicle into place for digging.
C. Operate dragline for digging.
  1) Lower bucket to digging level.
  2) Scoop bucket in with overburden soil or mineral material.
  3) Raise bucket and move to sump area or hauler.
  4) Dump bucket and return it to digging area.
  5) Repeat steps C. (1) through (4) above.
D. Shutdown in case of emergency failure.
  Note: An audio voice signal should announce the nature of the
    emergency—excessive engine temperature, a tangled drag-
    line, etc.
E. Reset circuits following emergency shutdown.
F. Adjust for nighttime or low-light level operation.
  1)Turn on internal lighting.
  2) Turn on floodlighting.
G. Adjust cabin temperature if required.
H. Operate windshield defroster, washer and wiper as needed.
 I. Communicate throughout for coordinating safe dragline operation:
  Pitman
  Standby maintenance man
  Hauler
  Main office

6) Control-Display Requirements

Following an activity analysis, a designer can begin to select or design appropriate controls and displays. Human factors principles of

design should be incorporated throughout.[3] This specification of controls and displays following an activities breakdown is referred to as a "coupling-mode" analysis or the interfacing of the HO with the machine. Each activity must be examined in detail to establish human factors criteria to be used in design. The dragline example is outlined in Table 8.1 and is illustrated in Figure 8.1.

7) Layout and Environmental Protection Requirements

The arrangement and spatial layout of the controls and displays at the HO's dragline workstation depend upon the anthropometric design strategy and criteria adopted by a designer (refer to Chapter 6). Obviously, the workstation must be designed to accommodate whatever the sizes might be of operators in the vehicle. If only a few HOs are to be scheduled to operate it, the vehicle workstation can be pretty much tailored to them. However, over a period of time, the dragline will most likely be operated by numerous different HOs of various body sizes. It then follows that a range of different body sizes should be accommodated in designing the workstation. This can be done by designing the console height for an average-size HO. Seat adjustability can be provided for raising small operators and moving them within easy reach and visual access of the dragline controls and displays. Control reach, for example, should be within 27 inches from the seat reference point (SRP) to accommodate a 5th percentile male HO, or it should be about 25 inches from the SRP to accommodate a 5th percentile adult female.

To provide direct visual contact in controlling the dragline bucket, a wide visual area must be provided, with the workstation located to one side to eliminate visual obstruction by the boom structure (see Figure 8.2). The seat adjustment should provide for a comfortable arm angle at the grip stick with arm and elbow support for precision control. The required force actuation in setting the mode lever and moving the vehicle-and-dragline grip lever should not exceed the capabilities of the 5th percentile male or the 5th percentile female if females are scheduled to operate the rig (refer to Chapter 3).

Environmental protection requirements must, of course, also be determined along with the activity requirements. These include provisions for

---

3. Refer to Chapter 5 for guidance documents.

Figure 8.1. Dragline control-display HO interface design.

Activity	Control Description	Associated Display	Console Location	Reasons
A. Startup.	Turnkey ignition switch with "Off-On-Start" positions.	Controls energized: engine sounds.	Left center.	In keeping with automobile ignition habits.
Vehicle control (also dragline control).	Grip control lever. (see Figure 8.1). Forward movement drives vehicle at maximum speed of 3 MPH; rearward brakes vehicle; left turns left; right turns right.	Vehicle scene of outside shifts in vehicle movement.	Center at elbow level, operable by either hand.	Multiple-dimension tracking is required. Single lever simplifies control. Control movement directions are in keeping with those of the vehicle.
B. Lock vehicle in place.	Springloaded three-position selector lever. (see Figure 8.1). In "Vehicle Drive" position, the vehicle lock is released and the grip control lever operates in vehicle control. In "Dragline" position the vehicle is	Lever position reading	Left side, elbow level	Operation is simplified with a single-mode switch/lever that permits the use of one control (the grip lever) for two different complex functions. Location to the left is out of the way after setting.

locked into position for digging, and the lever controls the dragline. The lever is detented, snapping into position.

| C. Dragline Control (also vehicle control). | Grip control lever with momentary thumb-actuated slideswitch at top (see Figure 8.1). Forward movement pays out dragline with amount of dragline proportional to lever displacement (one foot per second maximum rate of bucket movement). Rearward, pulls dragline in. Left, moves boom to left. Right, moves boom to right. Forward on thumbswitch lowers bucket. Rearward on thumbswitch raises bucket. | Direct view of bucket action. (Cabin is designed off-center to permit unobstructed view of bucket and lines.) | Refer to "A. Vehicle control" above.

Multiple-dimension tracking is required. A single dual function control simplifies overall control design. Control movement is in keeping with direct movements of the dragline. The use of the same control for vehicle and dragline is also easily discriminated by the different visual scenes and body-motion sensing. The thumb-actuated slideswitch directional control is also consistent with bucket action. |

125

Table 8.1. (*continued*)

Activity	Control Description	Associated Display	Console Location	Reasons
D. Emergency shutdown.	Recessed pushbutton outlined in red.	An auditory voice signal calls out on the speaker the nature of the vehicle or dragline emergency, which is dependent upon how the emergency conditions are instrumented.	Control is located at the right side forward area at elbow level. Signal is from the speaker at front.	Fast emergency action is provided with a pushbutton. An auditory signal is used for an emergency since it intrudes most effectively on the HO's attention. A voice signal is used to provide immediate intelligence concerning the emergency. Asignal cutoff is necessary to eliminate the noise after the HO is notified.
E. Resetting after emergency shutdown.	Recessed pushbutton outlined in yellow.	Amber, push-to-test "Reset" indicator light.	Right side forward area.	Pushbutton allows rapid resetting. Yellow outline indicates that caution should be exercised in resetting. The indicator light reminds the HO that emergency disabling of

				circuits requires re-setting. Recessing of pushbuttons makes it less likely that they will be hit inadvertently.
F. Turning on **console** lights.	Rotary knob with "Off" position and increasing brightness CW.	Panels light up.	Left side at elbow level.	Provides easy access in dark.
F. Turning on cabin lights.	Two-position toggle switch.	Cabin lights up.	Left side at elbow level.	Provides easy access in dark.
F. Turning on floodlight.	Two-position toggle switch.	Outside lights come on. An amber push-to-test indicator light comes on.	Right side, high.	Separated from other lighting controls because lights are located outside. Indicator light tells the HO that lights have been left on, which might not be evident in daylight.
G. Adjusting cabin temperature	Linear lever off at center, back for air conditioning and forward for heat.	Lever setting position.	Right side, elbow level.	In keeping with automobile position habits.

Table 8.1. (*continued*)

Activity	Control Description	Associated Display	Console Location	Reasons
H. Turning on defroster.	Two-position toggle switch.	Switch position.	Left center at elbow level.	Provides quick access.
H. Turning on windshield washer fluid.	Pushbutton, momentary.	Fluid hits the windshield.	Left center.	Provides quick access.
H. Turning on windshield wiper.	Two-position toggle switch.	Switch position; wiper action.	Left center.	Provides quick access.
I. Selecting the office or the dragline area for communication.	Three-position pointer selector knob.	Pointer knob position.	Right center just below the speaker-microphone unit.	Pointer knob shows the position setting most directly; location is conveniently accessed in communication.
I. Turning on intercommunication circuits.	Grip-stick mounted, springloaded momentary pushbutton for forefinger operation.	"Click" feel of pushbutton; auditory signals from speaker.	Grip-stick mounting of intercom button; microphone-speaker unit at right center.	Stick mounting provides quick access. The microphone-speaker unit location is within easy voice range of the HO.

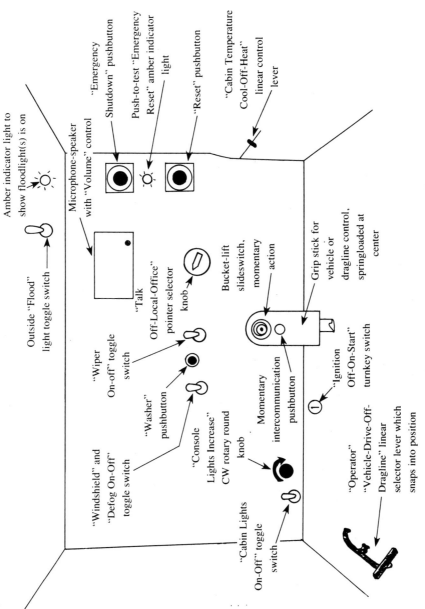

Figure 8.1. Dragline control-display HO interface design.

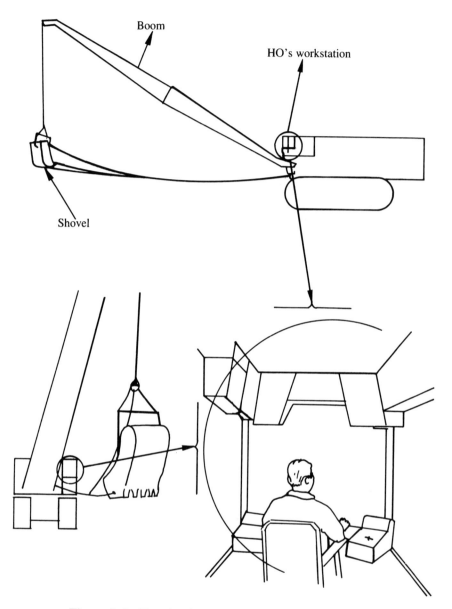

Figure 8.2. Sketch of a proposed dragline workstation.

comfortable temperatures and humidity, adequate inside-outside light levels, and for clearing away windshield dust raised during digging. Attenuating high noise levels put out by the dragline as well as damping induced vibrations must also be considered; the large areas of transparency at the HO's workstation, for example, may vibrate sympathetically through the boom structures and must be controlled with effective window-frame insulation and damping devices. Engine noise within the compartment must also be controlled so as not to exceed acceptable operating and comfort levels.

8) Testing Prior to Fabrication

Before the dragline design is firmed up, several types of tests should be completed to assure the workability of the design. These might include the following:

A. *Fiberglass or wooden mockups*. These should provide a three-dimensional perspective for a designer. Designers often have difficulty visualizing a workstation's spatial arrangement simply from looking at drawings. For that reason, some sort of mockup may be mandatory. Fairly sophisticated mockup exercises should also be completed by sampling different subjects representing various HO body sizes. Reachability and knee, arm and shoulder clearances are some of the considerations to be tested. Control postures, operating sequences with a link analysis in mind, and various alternative control-display layout arrangements can be assessed. These can be given a fairly extensive critique in discussions with the different designers and a number of prospective dragline operators.

B. *Computer-based testing*. Computer facilities are, of course, required. Apart from computer-aided design techniques, which can provide various dimensional perspectives, computer tests can also involve simulation testing with the HO in the loop. A computer program can provide an analogue of the draglift device by simulating the boom, draglines and bucket. Control movement rates versus dragline movement rates (i.e., the ratios) can be studied and optimized through such procedures. The lag times involved in dragline tracking can be tested as well to determine lag tolerances relative to required accuracies in positioning. Lags, for example, should not exceed 0.3 seconds from the time a control is moved until the device is actuated.

The use of computer dragline simulation techniques can also serve a training function. Indeed, simulation can be quite useful in helping HOs acquire a measure of tracking skill with the dragline well in advance of the actual availability of operational or prototype equipment.

C. *Building a prototype model.* Sooner or later, a prototype dragline vehicle will have to be built. The first model can then be tested with the HO in the loop to see if all the controls and displays and their layout arrangements are functional, and that the job can be done efficiently with the equipment provided. Several different operators, preferably of different body sizes, should be permitted to operate the rig, and they should then be extensively debriefed to obtain as much information as possible for improving the design.

Summary

Procedures described in this chapter can be applied in the human factors analysis of newly-designed products. These are applicable whether the product is simplex or complex. A simple analytical approach should include the following:

1) A statement of requirements and purpose.
2) A preliminary design study.
3) A means selection.
4) A specification of design requirements.
5) An activity analysis.
6) A control and display derivation based on activities.
7) An operational environment analysis to include the workstation layout and provisions for safety and protection.
8) Workability-assurance test procedures before fabrication of the final product.

An example is provided in the human factors analysis of dragline digging equipment.

Exercise 3

Now it's *your* turn!

In the previous exercise completed in conjunction with the material of Chapter 7, you chose a product currently on the market and attempted to improve it through the application of sound human engineering principles in redesign.

Now, begin with a *proposed* product that has *not* as yet been designed. Select any simplex or complex product you like. Remember, though, that you must start as though it has never before been designed so that you can apply all the human factors design procedures described above.[4]

Here are steps to consider:

1) State the unique purpose of the product.

2) Complete a preliminary design study. (This should be sufficient to demonstrate that you know how to go about it.)

3) Select the means by which the purpose is to be accomplished. (Explain why this means of meeting the unique purpose of the product was selected rather than some other. For example, why is an electric hair dryer to be used rather than a towel to dry hair?)

4) After the means selection, outline the design specifications for the product. (For example, having selected an electric hair dryer rather than a towel, it must then have certain safety features, be lightweight for easy handling, etc.)

5) Outline all necessary activities for using the product. Then, specify the controls and displays needed to accomplish these, together with the design considerations important in laying them out and for providing environmental protection. Provide a rough sketch of the proposed configuration.

6) Describe what you would do to test the suitability of the product in serving the intended purpose stated in number 1 above. If you have trouble getting started, it would be helpful to discuss your project idea in class.

---

4. Regardless of how many of the type of product you select that are already on the market, e.g., radios, hair dryers, washing machines, etc., for purposes of this project you must assume that it has never been designed before and therefore must start from scratch.

VI
Total Systems

9

Human Factors Applications in the Design of Complex Systems

In the last two chapters, human factors applications were discussed with respect to simplex and complex products. Chapter 7 treated the case of product improvement where a product was already on the market. In Chapter 8, a product was considered for human factors applications when it was being newly developed. In both cases, the products were largely considered as single entities—appliances or equipment used in an independent context by the consuming public. In the case of complex systems, however, the total operational complex with its interdependencies of equipment, facilities and HO personnel has to be considered in the development process.

The Nature of Systems

A system embodies complex interactions which can be broken down into subsystems and components for an analysis of their interdependencies. An automobile, for example, can be analyzed as being made up of such subsystems as the roadway, service stations, traffic regulations, signage, etc., as well as the vehicle itself which has engine coolant, lubrication, fuel, a control-display console, a passenger compartment, and other subsystems.

In the total perspective of things, the operation of almost any product or piece of equipment can be considered as an interdependent entity that can be analyzed by its subsystems and components. Thus, even such a simplex product as a hair dryer is part of a more total system of personal cleaning and grooming.

Stages of Development

A systems approach, in which human factors principles can be applied, can be undertaken at any time during the history of a system. A system such as the criminal justice system or an international airport passenger-processing system—i.e., those that have been operating with the same essential design for decades—can still be analyzed for human factors improvements. Other systems, such as the Strategic Defense Initiative, popularly called "Star Wars," may be at only an abstract-idea level for future operational developments. They can still be fully analyzed, however, as they proceed with developments in incorporating detailed human factors principles in design applications.

System Goals and Measurements

Regardless of at what stage of development or operation a system exists, further development or improvement goals can still be formulated. In an automobile vehicle system, one that has been operating for over 90 years, such improvement goals can be stated as reducing the current accident rate by 20 percent within the next decade. Current airport antihijacking system goals could involve a further reduction in the number of successful hijacking incidents over the next year. An operating computer system might involve such goals as an increase in productivity by five percent and a reduction in operating costs by 10 percent during the next fiscal period.

Systems being developed for future operations can, likewise, be related to overall goal statements. Requirements for new transit systems to be developed can be stipulated in such terms as, "300 passengers to be moved over a 45-mile urban space per hour for each unit of travel." A weapons system to be developed can be goal-directed in terms of payload, distance and strike accuracy requirements, etc. well in advance of any of the hardware that is yet to be designed.

The Power of the Systems Approach

The dynamic potency and effectiveness of a systems approach in development resides in the scope of system elements that are being considered. All of these elements can be directly related in analysis to the accomplishment of total system goals as illustrated in the following examples.

The Automobile

In striving to reduce accident rates, many facets of the automobile's operations can be studied from a systems point of view. Operations Research (OR) studies, for example, have indicated that a greater rate of accidents occurs at intersections than on highways. Thus, law enforcement at intersections is more effective in preventing accidents than is the enforcement of speed laws.

Several other areas of control can also be shown to reduce accidents through redesign of the vehicle itself. Introducing the "stop lamp," for example, in the rear window beginning with 1986 models is expected to reduce the number of rear-end collisions. Studies by the Motor Vehicle Manufacturers' Association have shown that drivers tend to look through the windows of the cars in front to see what is farther down the road. They are thus more likely to see the brake light up ahead when it is in the rear window than when it is at bumper level (Sivak, et al., 1981).

The use of seat belts is also a vehicle-design area that can be further developed in reducing accident rates involving personal injury. Almost every person, according to the Automotive Information Council, will be involved in an auto accident during his or her lifetime. Half will suffer a disabling injury, and one in 50 will be killed, according to current statistics. Only 15 percent of all motorists now wear seat belts, even though studies have shown that seat belts really do reduce injuries and fatalities. This area of vehicle design could, thus, also help to accomplish the total system goal of reducing accidents.

Other studies have shown that more than half of all traffic deaths occur at night. The rate is three to four times as great as during the day (Leibowitz and Owens, 1977; Mishlin, et al., 1983; Leibowitz and Owen, 1986). Apart from fatigue and alcohol abuse (Ranney and Gawron, 1986), the cause can be shown to be reduced visual capacity in low-level light. Improvements in headlight design and driver training for nighttime visual limitations could help reduce accidents. In this case, design improvements across a number of the total subsystems could also help to reduce overall accident rates, e.g., improved roadway lighting, traffic-control subsystem improvements through the mandatory use of reflectors on vehicles and more extensive nighttime speed laws, etc.

Airport Antihijacking Security

Reducing aircraft hijacking rates can also best be accomplished by relating to the system across the board rather than just to one or two of the subsystems. All the subsystems and various operational facilities must become involved. Hijackers clandestinely board aircraft with weapons or explosives and subsequently take hostages, thereby forcing the crew and passengers to do, under threat, whatever they are told to do. Demands are also made to outside establishments under the threat of injury and death to those held captive.

Antihijacking security primarily concerns the prevention of weapons being secreted aboard aircraft. This not only involves carry-on baggage inspection, but every facet of the operation where weapons can be smuggled aboard—during the on-board cleaning, maintenance and servicing of the aircraft, during baggage-compartment loading and un-loading, and so on. Indeed, any access to the aircraft—when on an unsecured parking ramp, when adjacent aircraft are being boarded, etc.—must be considered. Sound human factors principles of design should be applied across the board and every facet of the system considered if the system goal is to be accomplished. This means that the total system and all of its subsystems where vulnerability is at risk must be considered if the full power of the systems approach is to be utilized.

Computer System Parameter Control

A computer is the classic example of a system application. However, even here, the comprehensive power of a systems approach is sometimes overlooked. For example, a major system parameter is the overall cost of operation. Operations research studies have indicated that nearly half of a total computer budget goes for software development or the programming function (O'Hare, 1981). In meeting a total system objective of reducing costs, this programming activity should thus be carefully studied for human engineering applications, since this is an area where major cost reductions could be realized.

Analytical Techniques

In both active and ongoing systems and those to be newly developed, similar analytical approaches can be applied. General system requirements, of course, must first be spelled out, and the required total systems performance criteria specified. Some analyses that can be undertaken in a human factors systems study include:

1) An outline of all subsystems.

2) The specification of requirements for each of these.

3) The development of a mission profile.

4) A functions analysis.

5) An allocation of functions for human operator or machine performance.

6) An activity or task analysis, depending upon whether improvement of an old system or development of a new one is involved.

7) An information-requirements analysis.

8) A coupling-mode or interface-design analysis.

9) An analysis of environmental conditions and hazards that will be involved.

10) An analysis of job aids required.

11) An analysis of personnel skills and training that will be required.

12) The development of a plan for maintainability.

13) Preparing and conducting a system organization and testing program.

1) Outlining the Subsystems

Total systems are comprised of interdependent subsystems that are made up of human operators, machines, facilities and equipment. A system of automobiles, for example, includes the following subsystems:

- The vehicles
- The drivers, their skills, and training programs
- The roadways, their routing, construction and materials used, and their maintenance
- The traffic regulation, enforcement and control devices
- Vehicle service and maintenance stations
- Accident insurance and litigation processes

A computer system, as another example, consists of such subsystems as follow:

- Programming languages and techniques
- Programmers, their skills, and training programs
- Conversion equipment and procedures that convert programs to machine-processible form (keypunch, discs, tapes, video-display terminals, etc.)
- Central processing equipment and circuit modules with memory and control
- Output equipment, procedures, and training (high-speed printers, voice-output and reaction equipment, video-display terminals, etc.)
- Facilities and environment-control procedures and equipment
- Output facilities for robotic computer-aided manufacturing, computer-aided design, and various other computer-aided work applications

An airport antihijacking security system is still another example of a total system that consists of such subsystems as follow:

- Carry-on luggage and loaded-baggage inspection operations
- Passenger-boarding security checks
- Baggage handling, loading and security operations
- Cleaning personnel screening and on-board security operations
- Maintenance personnel screening and on-board security operations
- Parking ramp surveillance and security control

2) Preparing Subsystem Design Specifications

A preliminary design study of each subsystem will permit the development of requirements specifications for each subsystem. For example, a computer output-input video display terminal subsystem should be specified to be "glare-proof," "worker adjustable," "an activity that produces minimal fatigue," etc. An antihijacking airport security system might specify requirements for a carry-on luggage-inspection subsystem as follows:

- Privacy (minimal exposure of personal clothing and items to public view)
- Minimum inspection time (not to exceed 30 seconds per inspection by unit)

- Minimized handling (primarily handled by the passenger depositing baggage on and taking baggage from the conveyor)
- 100 percent reliability in the detection of weapons and explosives

Such requirements for each subsystem must be generally specified at the outset. Human factors design applications can then be used to meet these requirements.

3) Developing a System Mission Profile

To orient the human factors analysis directly to system performance, a broad perspective is first needed on just how the system operates. This can be achieved by outlining mission phases within which a more detailed functions analysis can proceed. A computer mission analysis might involve these phases:

- Program preparation
- Conversion to machine-processing form
- Central data processing
- Output-input interaction and data exchange

Such a profile then permits the analysis to proceed with operational details being further outlined within each of the phases above.

In the airport antihijacking security system, the following mission phases and segments can be identified:

- Aircraft maintenance security control
- Aircraft cleanup security control
- Aircraft fueling and servicing security
- Passenger baggage check-in security
- Baggage-loading-onto-aircraft security phase
- Carry-on luggage inspection phase
- Passenger boarding
- Inflight security surveillance phase

Each of these mission phases can then be subsequently analyzed for more specific functional details and human factors applications.

4) Performing a Functions Analysis

The mission profile, with an outline of operational phases, provides the framework for a functions analysis. This consists of a general listing of tasks that must be done within each mission phase, which is, at this time, without regard as to *how* they are to be done, i.e., by HOs, by a computer, by a servo-mechanical unit, etc. In the Program Preparation phase of a computer mission, for example, such functions can be identified as "program design," "program coding," "program debugging and program revision." Subfunctions can be further determined, such as "text preparation" during the design part, "flow diagramming and procedures implementation" during the program coding function, and "grammatical, syntactical and logical error detection" during the debugging and revision function.

Antihijacking security mission phases can likewise be broken down into their functions. The aircraft Maintenance Security phase, for example, would include such functions as maintenance-personnel security screening, aircraft access surveillance, post-maintenance aircraft security checkout, etc. The Cleaning, Fueling and Servicing phases would involve similar functions, which would be subsequently broken down into subfunctions. Personnel security screening, for example, would include such subfunctions as background reviews of individual personnel, searches of police and FBI files, etc. Post-maintenance aircraft security checkout would involve such subfunctions as passenger compartment weapons searches, luggage compartment weapons searches, and crew area searches.

The functions to be itemized for each mission phase can be determined on simply a logical and rational basis from the earlier iteration of system design requirements. Other approaches to a functions analysis include discussions with planners and engineers. At the outset, such functions can be considered to be merely those which have to be done in order to complete a successful mission.

5) Allocating System Functions

After functions have been listed, an early approximation can be made as to how they can be accomplished. The HO has certain capabilities and limitations as previously discussed in Chapters 2 and 3; these should be taken into consideration when designing functions for implementation. In the airport carry-on luggage inspection phase, the function involving a high rate of inspections can be done by direct visual means, viz., opening the case and viewing its contents. However, a 30-second time limit would be exceeded; thus, this function could not be totally allocated to the HO. If an X-ray machine were to be used,[1] the luggage could be placed under the machine by the HO. However, this too would be undesirable in that the HO would be exposed to X-ray radiation while, at the same time, failing to meet the requirement for a 30-second-per-unit inspection rate. The function allocation, in this case, points to a mechanical luggage-moving component such as a conveyor.

The 100 percent detection reliability requirement, on the other hand, may necessitate the use of well-trained hand searchers. Existing state-of-the-art mechanical searchers may simply not provide enough reliability.

Another example of functions allocations would be in the aircraft-access surveillance function in the aircraft maintenance security phase of the mission. An HO could serve as a walk-around guard to look for any "suspicious" incidents. Detection probability would be quite unlikely in such a design since the HO is a poor monitor of such events. Another design alternative would have the HO at a video observation station with the various aircraft compartments covered by video cameras. Since the HO is a poor monitor, this allocation would also be poor because the screens must be watched. A functions allocation, in this design, would provide for the machine design to "monitor" or alert the HO. For example, any action in a compartment would activate photoelectric cells to evoke a sounding mechanism to attract the HO's attention. The HO could then turn on the television camera for surveillance.

Each function must be carefully studied to make the best allocation decisions regarding the HO's capabilities and limitations and with regard to how the machines can best be designed to accommodate these. Throughout the course of development and design, the functions allocations must be largely subject to change as events may show that previous tradeoff decisions might not be effective, i.e., on the basis of observed performance, as well as projected costs (Chapanis, 1965; Edwards and Lee, 1972; Price, 1985).

---

1. With the development of plastic handguns and explosives, metal-detection equipment may be inadequate, while X-ray capability requires further study. Vapor "sniffers" may be more efficient in detecting explosive substances and detonators, but these also require further study and development. New equipment developments, such as the Magnetic Resolution Imaging machine, should also be considered (Cranston, 1986).

6) Completing an Activity/Task Analysis

After the functions have been clarified and allocation decisions made, the designer can then proceed with an activity or task analysis. These types of analyses were previously described in Chapters 7 and 8. In Chapter 7, the task analysis procedure was outlined as employed for a product-improvement review. In the case of a task analysis, there are specific controls and displays that can be analyzed for their adequacy in "feel," position reading, information feedback, and so on. This was illustrated for such specific current market items as a Hotpoint clothes washer and a Bel Tronics radar detector. In a systems-analysis approach, a complete array of equipment and personnel already operational could be studied for improvement by means of the task analysis method. Another case for using the task analysis method, of course, is when simulation is being developed and it becomes necessary to know control-display action characteristics in detail in order to replicate them for research or training purposes.

The activity analysis method was described in Chapter 8 for a typical case where new equipment is being developed. The list of functions provides the framework within which the activity analysis can proceed. For example, in the airport security function for inspecting passenger carry-on luggage, the equipment must be turned on, e.g., a conveyor, a "sniffer," an X-ray unit, etc. This is a discrete action. Others—such as following a piece of luggage while checking for weapons—are tracking activities.

7) Completing an Information-Requirements Analysis

The determination of information requirements can be considered to be analytical subprocedures within the activity analysis. The different types of display information were described in Chapter 5 within the context of each activity. To reiterate, check displays are used where only discrete information is required, such as when power is on or off; the equipment is thus either in a go condition or it is not. Indicator lights and audio signals are most frequently used for check displays. Indeed, such simple displays are preferable to more complex ones whenever possible in order to simplify the HO's reading requirements. However, go-indicator displays are often unnecessary when other operating conditions provide the indications themselves, e.g., motor noise.

Qualitative information displays show a magnitude and direction of change in a condition. The speedometer in an automobile and a barometer that shows when pressure is rising or falling are qualitative displays.

If reading specific quantities becomes necessary, the best type of display is a counter or digital readout. Quantitative readings take up more of the HO's operating time and so should be avoided as much as possible. For example, in driving an automobile, the speedometer reading is most effective for an increasing or decreasing indicant within a given range. A precise quantitative reading seems to be unnecessary here and undesirable from a reading-time standpoint.

Just as with the control task/activity, the reading task/activity should be kept as simple as possible. Qualitative readings are preferable to quantitative ones, and check readings are preferable to qualitative ones.

8) Designing the Interface

This design procedure was illustrated in Table 8.1 of Chapter 8 for the dragline vehicle design. The control-display coupling of the HO with the machine is accomplished directly through the activity and information-requirements analysis. For example, the passenger carry-on baggage security inspection could combine operations into a single action; a two-position toggle switch might be used for jointly turning on the conveyor, the sniffer mechanism, and the X-ray machine. The conveyor motor noise and belt movement would be sufficient feedback signalling the adequacy of the control response. Energizing the display screen may also be a sufficient indicator to show that X-ray and vapor surveillance is operating. However, an amber light signal should be provided with a caution note to stay clear of harmful X-ray radiation.

A discrete signal display—such as an auditory "beep"—should call the HO's attention to anything of a suspicious nature in a bag for closer examination on the X-ray machine or the sniffer mechanism, or for direct visual examination as needed. This discrete signal would be a preferable reading to that of a qualitative pointer dial or a numerical counter of sniffer units from the standpoint of processing time and actions required in response.

The interface design would also include anthropometric considerations. Designing for HOs and the general passenger population in carry-on inspection, for example, would require accommodating various body sizes. In a sitting-standing workstation, a glare-free screen, angled for proper viewing by all sizes of security agents, must be provided. Passengers, regardless of their body dimensions, must be able to position their baggage on the conveyor for inspection.

Another important design consideration is the layout of the inspection station with respect to the total antihijacking security system. It could, for example, be located at a common port of entry to all passenger boarding areas, or at individual flight lines. Such system layout design decisions must be made on a cost-effective basis to meet overall security objectives.

9) Considering Environmental Conditions in Design

The environmental conditions under which an HO must operate should be analyzed to assure that adequate protection is provided and that the required performance levels will not be adversely affected. A machine might put out harmful fumes or radiation levels against which an HO might have to be protected. The operating environment may require that conditions be controlled to assure proficient HO performances, viz., temperature and humidity, light, noise, vibrations, etc. At the passenger carry-on baggage security inspection subsystem, X-ray machines may require lead screening to prevent the diffusion of harmful radiation. For direct visual inspection, 200 to 300 foot-candles of illumination will be required for the detailed discrimination of possible weapons or explosives. Danger to the security agents from potential hijackers must also be considered, with the suitable stationing of airport police, etc. The conveyor environment must also be adequately guarded to, for example, prevent passengers or operators from becoming entangled in any exposed moving parts.

10) Developing Job Aids

Each job or position within a system should be studied to determine specific aids that would assist an HO in doing the required job functions. These might involve operating instructions embossed on placards and attached to the machines, appropriate caution and warning signs, making instruction cards and manuals available, etc. In the antihijacking security system, in addition to operating aids at each HO station, an operating manual could be provided that would instruct all operating personnel on what to look for in a potential hijacker, where the danger points of weapons transfer may be, and what suspicious actions should be noted for further observations or close inspections.

11) Developing a Training Program

Simple and complex systems require that the HOs be given some degree of training or instruction. In some cases, simple instruction pamphlets may suffice. In other cases, more complex systems-involved training procedures may become necessary, ranging from simple on-the-job training programs to classroom instruction supplemented with practice on part-task trainers or detailed whole-task trainers and complex simulators. The required training functions must be determined throughout the system, not only to determine the skills and training necessary for the HOs to operate the equipment, but for maintaining it as well.

12) Developing a Plan for Maintainability

Maintenance, though sometimes neglected as a system area, can be of fundamental importance to system efficiency (Goldman and Slattery, 1964). Many systems, in fact, because of their highly sophisticated nature, are prone to malfunctions simply due to their innovative characters and sheer complexity. Without adequate maintenance planning, their effectiveness can be significantly compromised.

A distinction can be made between maintenance and maintainability. Maintenance comprises all actions necessary to assure reliability in preventing equipment breakdown and restoring it to operational readiness when it does. Maintainability is simply the ease and efficiency with which the maintenance activities can be conducted.

Maintenance includes such functions as fueling and lubricating, or what is called preventive and scheduled prefailure maintenance, and repairs at the various levels or echelons of service. In the maintenance plan, down time—the period during which the system is inoperative—must be minimized.

A preliminary maintenance plan should involve the following:

1) The determination of servicing and overhaul requirements.

2) Establishing the failure probabilities in electrical, hydraulic, electronic, coolant, fueling, mechanical, etc., components.

3) The determination of the levels or echelons of maintenance. For example, in the first echelon, the HO may simply change a tire or remove stuck paper from a copying machine; the second echelon is at a technician's level, such as is done at a service station; the third echelon is at the depot level. Here, the removal, repair and replacement of malfunctioning equipment are carried out by trained and skilled maintenance personnel who also perform preventive routines. The latter are exemplified by garage, shop or factory facilities.

Good maintainability design requires that the designer include the following in a maintainance plan:

1) Reducing the number of complex skill requirements to a minimum through the design of simplified maintenance procedures. This can be accomplished, for example, by the use of throwaway units for replacement and with capsular assemblies designed to be easily removed and replaced.

2) Designing for skills that are readily available in the general population. Maintenance skills in mechanics, electronics, hydraulics, etc. are commonly available and should be sufficient for any new maintenance job with only a minimum of instruction.

Maintainability Design Features

To ensure that efficient maintenance behaviors are employed in the system, effective rules should be applied in design as follow:

Labeling

1) Establish that the meaning of a label is most significant to the maintainers, using only words and symbols with which they are familiar.

2) Employ labels in consistent locations on the equipment, i.e., always above, always below, etc. Make them durable and easily readable.

3) Label connectors by the type of input or output, and locate the labels next to the aperture, receptacle or connector.

4) Design instruction plates to be easily read, to be brief, and which are succinctly diagrammed.

5) Make danger and warning signs white on a red background. Use only horizontal lettering. Make signs conspicuous and brief.

Accessibility

1) Design adequate workspaces where crawling or kneeling is required. One-hand and two-hand access should be adequate according to the task requirements.

2) Locate the most frequently failing components most accessibly.

3) Design panels that can be removed with common tools or finger-actuated screws.

4) Design an item weight to be less than 45 pounds if it is to be removed; if difficult to reach, less than 25 pounds.

5) Design filters and strainers to be easily accessible.

Repair, Removal and Replacement

1) Design modular or unit packaging.

2) Design self-lubricating and sealed assemblies as throwaways.

3) Design storage locations and parts which have clear markings.

4) Design snap-in, quick-disconnect assemblies.

5) Design for the use of only common handtools when possible.

6) Design for foolproof connections, with specific guide pins, keys, and prongs and interconnectors of different sizes and shapes.

7) Design equipment covers with handles that are well-insulated when surfaces are hot.

8) Design covers with hinges, latches, and catches to reduce handling.

Inspection and Testing

1) Design to minimize special test equipment.

2) Design for reservoirs, gauges and meters to be easily seen without the removal of panels.

3) Design seals and gaskets for easy viewing after installation to allow for easy checking.

Troubleshooting

Troubleshooting is simply the procedure of finding out what is wrong with a product when it will not work. Troubleshooting can be facilitated by:

1) Including specific symptom-malfunction data in the maintenance manual.

2) Include step-by-step procedures, spelling out how to locate the trouble area.

3) As a design procedure, insist that the manual writers themselves execute the troubleshooting procedures to verify that what they have written is accurate, precise and understandable.

13) Organizing and Testing the System

The systems design process must be an iterative one in which study and test results are continually being fed back to the designers. In early studies, system components are diagrammed and flowcharted. Preliminary time-line studies are completed using the best guesstimates of task durations and time-sharing sequences that can be linked together.

Early layout studies can be completed utilizing 5th and 95th percentiles of "to-scale" manikins along with drawings. Workstation mockups can then be completed from isometric drawings and three-dimensional cardboard models. Later, wood and metal mockups can be employed for a wider participation of designers and users in the evaluation process.

Where difficult work sequences and situations are involved, simulation exercises can be worked out. For example, driving simulators are often used to test steering control ratios and the effectiveness of road signage.

As the system advances to its hardware stage, prototype equipment can be tested for operational feasibility. Final design configurations can later be employed to test the overall system before extensive manufacturing of the system and its operational deployment are made.

Various environmental tests and measurements can also be made in developing design data. Human operator physiological and perceptual-motor behaviors can also be measured for systems performance. These will be discussed at greater length in the next chapter.

Summary

The systems design approach can be a powerful one when applied to accomplish goal-directed outcomes. Its effectiveness lies in the potential for design manipulation across the various subsystems to accomplish the desired outcomes.

System examples are cited, such as the automobile, the computer, and the airport antihijacking security system. Analytical procedures consist of determining the subsystems and their operational requirements and specifications. Then, hypothetical mission outlines are prepared along with functions analyses. The functions are characterized by human or machine involvement or for combinations of these according to performance criteria. From the designated HO functions, activities and their sequences are detailed, together with the HO's information requirements for operational readings in controlling and managing the system. These latter data then provide the basis for control-display design, while an environmental analysis can reveal the special protective and performance-aid provisions that will be needed. Job performance aids and training requirements can be specified after completing all the foregoing analyses.

The maintainability aspects of a systems design program are discussed as being highly relevant to maximizing operational efficiency. General system organizational and testing procedures are also described.

Exercise 4—Developing an Overview of the Systems Approach

The purpose of this exercise is simply to go through the analytical procedures as outlined in this chapter. For purposes of this exercise, please identify any complex operational system that might be analyzed by means of the systems approach, e.g., a supermarket, a recreational park system, a museum, a recreational boating system, a mass-transit system, a power plant, a dairy farm, etc. The system should be one of your own choosing.

Analytical Steps

Having identified an operation or system, state some quantitative goals for it. These are the ones toward which you should design. Then proceed as follows:

1) Outline the subsystems.
2) Prepare the subsystem requirement specifications.
3) Develop a mission profile.
4) Derive the general functions from the mission profile.
5) Allocate the functions, giving rationales for an HO-machine division of labor.

6) Complete a task or activity analysis for the HO.
7) Complete an information-requirements analysis.
8) Prepare a coupling mode analysis.
9) Identify environmental conditions and protective requirements.
10) Describe some job aids.
11) Describe personnel skills and the training required.
12) Discuss some maintainability aspects of the system.
13) Tell how you would test the system before making it operational.

Throughout the analysis, apply the human factors design principles you learned in Chapters 7 and 8. Prepare a brief report to show how you carried out the analysis.

VII
Design Evaluation

10
Human Factors
Evaluation Techniques

The human factors quality in a product or system can be assessed at any phase in the design process. Human factors data and design standards must first be incorporated in the preliminary design study. From that point on, various evaluation techniques can be employed to assure that good human engineering has been practiced in design.[1]

Early in the development program, flowcharts and diagrams can be used as design aids. In some cases, computer-aided anthropometric design programs can be used for sizing the equipment. As much as possible, drawings should be accompanied by documents substantiating the human factors criteria that have been initially applied in design. When design is so documented, it will assure that the original design was *not* arbitrary, and that it should not be arbitrarily changed without consideration of the original design intent.

Drawing Review

As soon as preliminary drawings have been completed, the early human factors aspects of design can be assessed. For example, the adequacy of the anthropometric design can be evaluated by the use of artic-

---

1. Evaluation is usually based on the manipulation of "independent variables" to assess their effect on a "dependent variable." The dependent variable might be operator performance as measured by time, error, etc. Independent variables can be changed, such as lighting, layout, etc., to determine how the dependent variable is effected.

ulated manikins that are of the same scale as the drawings. A 95th percentile manikin can be manipulated against the drawn images to see if there is sufficient room for a large HO to move about or if entrances and exits are large enough for easy ingress and egress; or if a 5th percentile female manikin can be appropriately postured to reach all controls and to see all displays, such as on a domestic appliance (Roebuck, et al., 1975).

The anthropometric dimension is, of course, only one aspect of human factors design. In studying the drawings before their release for fabrication, a fairly comprehensive human factors assessment should be made by means of an item-by-item checklist evaluation. Checklists should be more or less tailored to the specific product or system in evaluating both the operational and maintenance aspects of design. A designer can refer to the various human factors guidelines and standards cited in Chapter 4 for preparing checklists.

In this early phase, a preliminary human factors assessment can thus be made of operating controls and displays and their layout as well as of the complete maintenance aspects of the design.

Mockups

Before the design proceeds into the implementation of prototype models, simple mockups should first be made in order to carry out preliminary testing. In fact, mockups can be used quite extensively for effective preliminary assessments of almost any product or system component. In packaging design, for example, mockups or models can be used for color and configurational impact studies. Prior to making a packaging design operational, it might be almost mandatory to complete preliminary mockup studies.

Mockup studies are almost routinely done by many designers, who are then able to define specific problem areas more precisely than they would by simply trying to think them through in mental exercises. Milton Rosenthal (1973) described a mockup used at the Lockheed Company in Sunnyvale, California, to study a critical problem area in a microwelding operation. The operations involved welding and trimming delicate wire grid subassemblies. Workers complained of excessive fatigue and strain in the operational layout of their workstation. In redesigning the workstation, the industrial designers constructed a foam-core mockup. Dimensions designed for the new layout were built directly into the mockup. The designers considered the mockup to be an invaluable aid in verifying and correcting the new layout dimensions. For instance, the mockup served to point out potential pitfalls in clearances associated with elbow rests and adjacent cabinet drawers.

Apart from such rather routine applications of mockups in design developments, Carolyn Rozier (1977) has described a somewhat unusual use of the mockup technique. She studied the movement envelopes of amputee cases to establish workspace design parameters. The mockup consisted of a chair with a footrest, and frontal and side-grid areas as the mocked-up work surfaces. From the various movements of amputees in the mockup, she was able to determine that below-elbow amputees had an average decrease in workspace capability of 45 percent. Those with above-elbow amputations had an average decrease of 83 percent. Amputee workspace design developments, through such exercises as these, were thus made possible by using a simple mockup construction.

Mockups may, at first, be made of simple cardboard or fiber board, then, perhaps, of plastic, masonite, plywood or wood. The mockup can be used to test various manipulations of the product and to determine the best of several alternative configurations. For preliminary assurance of the general workability of an HO's work compartment, for example, tests can be completed as outlined in Figure 10.1. Large numbers of representative users need not be employed in these preliminary studies; rather, representative users at the extremes of body sizes is a more cost-effective alternative. A large operator, who measures around the 95th percentile in such body dimensions as buttock-to-knee length and sitting height, should be used for the upper limits of body sizes. Likewise, a small operator, who measures at about the 5th percentile of functional reach and sitting height, would be best for the lower limits. If the large operator has ample clearances and is comfortably accommodated in the compartment space, then the space should be satisfactory for all smaller operators. When the small operator can see and reach all controls and displays, then, likewise, all larger HOs should also be able to see and reach them as well.

Static mockups can also be used to test configurational adequacy in various manipulations. For example, a proposed camera design can be tested to see if "natural" thumb or finger movements might tend to block the lens. Two-hand or foot-hand coordinated activity requirements can be assessed for control layouts by having the HO run through the control

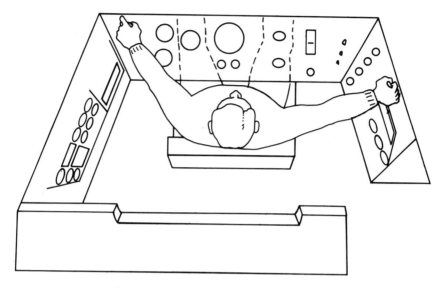

Tests:

1) Freedom of movement with personal equipment encumbrances (95th percentile).
2) Clearances and incidents of inadvertent bumping of controls (95th percentile).
3) Reachability of all controls under constraints (5th percentile).
4) Readability of all displays (5th percentile).
5) Performance of coordinated control-display sequences (5th and 95th percentiles).

Figure 10.1. Mockup tests of compartment dimensions and layout.

motions that will be required during operation. In this early phase of design, significant design faults can thus be readily identified through the use of three-dimensional mockups, while these might not become readily apparent simply from a review of the drawings.

Mockup studies should include a documented rationale for the control-display design and layout. During the critique of the design, explanations should not be necessary, while proposed changes should also be provided with rationales to prevent confusion.

160

Timeline Analytical Assessment Techniques

A good human factors analytical assessment technique also involves preliminary tests of feasibility through the development of operational timelines. A baseline of time can be derived from the operational requirements stipulated in the system specifications and projected through a series of mission sequences. A mass transit system must run on a precise time schedule, and industrial robotic operations require precise timelining of each function. These programming schedules can be used as the time base against which HO time functions on an activity-by-activity basis can be projected. Times can be measured or best estimates derived from whatever data are available. The timeline can be in minutes or, for more precise analysis, in seconds. The HO's measured operating times (if possible, derived in the actual design configuration) must then be shown to be feasible along the projected timeline of the mission. If not, the design configuration must be changed until the feasibility of required operating times can be demonstrated.

Information that can be obtained from timeline studies includes:

- Workload data for manning estimates
- Work areas where further automation may be necessary
- Critical points in a mission where stress is likely, especially in emergencies where time limitations can be critical

Assessment Through Simulation

Simulation is any means by which actual operations can be duplicated in order to study what the HO's reactions will be during real missions. It is a kind of animation of the mockup that can be used to test actual HO performance. For example, if an automobile driver's station is mocked-up using outside visual scenes, then dynamic simulation is involved. The visual scenes typically respond as movement-rate functions from simulated control input. Of course, the responses have to be of the same character and proportion as in the actual control situation to be a valid simulation.

Unfortunately, dynamic simulation can be very expensive. It can, however, be of great value in the following ways:

1) In assessing second-by-second events in the HO's interactions

161

with the system and testing capability in what, to all intents and purposes, is real-life performance.

2) In determining control-display stereotypes.

3) In testing dangerous operational sequences before they can become accidents with the actual operating equipment.

4) In providing safe training for the HOs who must operate the system for the first time. In simulation, the situation is without danger to personnel or equipment during the initial learning period.

5) In deriving actual HO time and motion data to be included in timeline analyses.

Simulators used for training can result in major cost savings as well as in providing a safe training environment. In aircraft flight training, high operational costs are involved in actual flight; a training hour in a simulator costs much less. Moreover, the noise and other distractions that occur in actual flight can be eliminated in the simulator, or introduced as training variables, thus making the simulator environment a much more efficient one for training.

Detailed fidelity in simulation, on the other hand, when important to testing, can be very costly to implement. However, here it is most important only in the early design phases when high-fidelity simulation may pay off by showing where design changes would be of most value. Later in design, such changes may be less feasible to consider since the design will have been finalized, and so such high-fidelity simulation testing would be of little consequence.

Simulation is now more extensively used in domestic programs, such as in automobile control design studies and nuclear power plant operational safety and training programs. Thus, recognition of its value in training and research should become increasingly widespread.

Assessing Prototype Models

After a first version of the product or equipment has been fabricated, it may still not be too late nor costly to introduce some model changes, particularly if these prove to be more or less critical to its safe operation, e.g., introducing guards at sharp corners or about moving parts.

Product Improvement

After the first product model has been mass produced and marketed, user reaction can then be tested. The assessment of the product is then directed to possible future improvements. HOs can be queried after acquiring experience in both operations and maintenance.

Where only small numbers of HOs are involved, personal face-to-face interviews can be valuable. When large numbers of HOs are available, self-administered written questionnaires are most effective. Questions must be carefully prepared in both instances and oriented around operational and maintenance aspects of the product design. Group questionnaires are generally most effective when structured for responses to be statistically processed. Different forms of questioning can be considered when they are best suited for eliciting responses about different issues to be raised in product design improvement.[2] In preparing questionnaires, different types of items can be considered, such as semantic differentials, rank ordering preferences or problems, comparisons, matching, identifying the best and worst features, etc.

Environmental Measurements

Assessments of operational and maintenance environments are best conducted early in the design program. If it is found, for example, that an HO will require protective equipment that could interfere with performance, further accommodations in design will become necessary.

Some safe limits of environmental conditions are given in Appendix B. More detailed environmental specifications for safety standards can be found in documents provided by the Occupational Safety and Health Administration (OSHA), 1825 K Street, N.W., Washington, D.C., 20006.

Common environmental measuring equipment includes thermometers for temperature, hygrometers for relative humidity, anemometers for wind, and barometers for air pressure. Other environmental measuring equipment includes:[3]

---

2. Questionnaire design is discussed at length in J. Burgess' *Designing for Humans*, Princeton, New Jersey: Petrocelli Books, 1986, pp. 459-462.

3. The various measuring instruments can generally be obtained through vendors by consulting the Yellow Pages under "Industrial" and "Laboratory" equipment and supplies, etc., or such reference documents as *Best's Safety Directory*.

- Light measurement—light and luminance meters
- Noise measurement—sound level meters and noise analyzers
- Vibration and acceleration measurement—vibration meters and analyzers and accelerometers
- Gases measurement—sniffers, detector tubes, and bellow pumps for air collection

In some cases, the environmental conditions might simply be observed, with further testing unnecessary. A light level of 100 millilamberts (e.g., under a lamp with a 75 or 100 watt bulb two or three feet away), for example, is bright enough for viewing all the spectral colors and for providing fairly good visual detail. If visual tasks impose more extreme demands on the HO, then more precise measurements are required. Likewise, the intensity of a noise environment can be estimated by gauging the level of voice volume in communication as follows:

Voice Level	Distance (feet)	Decibel Level
Normal	10	<45
Very loud	10	ca. 55
Loud shouting	10	65
Shouting	2	75

Where liquid or gas hazards are at all suspected, however, laboratory samples and testing may be mandatory.

Physiological Measurements

In some cases, an assessment of HO performance can be correlated with physiological measurements (Wierwille and Connor, 1983). The most commonly used HO physiological measurements in work performance studies are those of electrical potential changes in the brain (electroencephalograph) and muscles (electromyograph). The pulse rate is also commonly used and is generally measured by electrocardiography (ECG). Body and skin temperatures are sometimes measured, using thermistors attached to the triceps, fingers, forehead, over the stomach or chest, etc. Eye-movement measuring techniques are also being used in human factors studies; these measurements are called electrooculograms.

Such physiological measurements are, of course, somewhat costly to instrument, while their practical value is also sometimes questioned.[4] The instrumentation also creates an encumbrance on the HO that can interfere with smooth performance.[5] For such reasons, the value of physiological studies in human factors assessment must be carefully weighed for its significance.

Other Observational Techniques

The evaluation of human factors aspects in product design can be further accomplished by photographic, video, and timing techniques. Simple time measurements, through the use of such instruments as stopwatches and electromechanical timers, can provide basic HO performance data, particularly when time and motion data are required for timeline studies. Video and photographic cameras can be most useful when:

1) The HOs cannot be encumbered with body-measuring equipment that interferes with freedom of movement.

2) Time data are needed on a precise frame-by-frame basis or for hundredths of a second, with a precision chronometer superimposed in the action frames.

3) Special problem areas must be carefully studied to see what the HO persistently does wrong, or where potential accidents can occur.

4) Social or multipersonnel interactions must be studied for layout improvements.

Packaging Evaluation

Engineering evaluation tests are most commonly applied to product packaging design, such as for testing shock-absorption and damage resistance. Human factors aspects of packaging should also be assessed when products are boxed or packaged for marketing, shipping, storage,

---

4. Studies in work physiology, of course, have been found to be most useful in designing tasks and work modules in industry.

5. Miniaturizing instruments and radio and telemetry equipment has been employed to minimize these effects.

etc.[6] The various uses of graphic symbology to convey messages and "feelings" about a product or establishment should also be considered in this evaluation.

Lifting and Handling

The grasping-surface handle structure should be close to the body of the lifter(s) to minimize the possibility of low-back stress (see Chapter 3). Component weights should never exceed the lifting/handling capability of the intended user population. This aspect can be readily assessed by the representative "weaker" members of the population.

Labels

Caution, warning and advisory notes designed into the package can also be readily assessed by presenting dummy or mockup packages to a number of representative users for interpretation and followup.

Symbology

Pictorial representations and artwork, in conveying product or establishment information, may best be assessed by means of reaction questionnaires. (These were previously discussed under the "Product Improvement" section of this chapter).

The artwork of company logos and product signatures, designed to evoke a desired "image" or response, may best be tested through group questionnaires. In testing logos, for example, the questionnaires might contain a series of semantic differentials to evaluate the "character" that a designer wishes to impress on the consumer, e.g., weak-strong, modern-old fashioned, rich-poor, old-new, etc. Through this means, a preliminary assessment and followup design evaluation can be accomplished for the graphic symbology of the packaging.

Summary

Various techniques can be employed in the evaluation of human factors for products and equipment. In early design developments, drawing

---

6. These include the manual-handling aspects of packaging design (refer to Drury, et al., 1982) as well as the graphic-symbolic aesthetic aspects.

reviews can be accomplished using tailored checklists. Mockups can then be employed to give a more detailed three-dimensional perspective. Dynamic simulation can be employed to test specific HO capabilities and operational design deficiencies. Prototype models can also be used to determine critical deficiency. Individual interviews and group questionnaires can be used for product-improvement studies. Environmental and physiological measurements may also be pertinent to product assessment. Packaging design should also be considered for evaluation.

VIII
Product Safety

11
Designing for Product Safety

Designers must be especially concerned about the health and safety aspects of the products they design, for obvious reasons. OSHA formulates and regulates safety standards to which managers must comply in industrial equipment and environments as noted in the previous chapter. In recent years, designers, manufacturers and distributors have also become responsible and legally liable for any hazards or compromises with safety that their products might impose on the consuming public.

The Consumer Product Safety Commission

A consumer study commission was first appointed by President Johnson in 1968, with a charter to protect consumers against product hazards. The commission found that 20 million injuries occurred every year which could be attributable to unsafe product design. In 1972, the U.S. Congress passed a bill called the "Consumer Product Safety Act (PL 92-573) which established a Consumer Product Safety Commission (CPSC) as a permanent body. It was commissioned to:

1) Protect consumers against any unreasonable risks to their safety.
2) Evaluate product risks.
3) Develop product-safety standards.
4) Promote prevention research on product hazards and accidents.

Product-Safety Litigation
Product-safety law suits are generally characterized by:

1) Court judgments assuming consumers are unaware of product defects and hazards, and that they are, indeed, unable to evaluate these themselves (Kircher, 1970; Busch, 1976).

2) Manufacturers and sellers being held responsible for hazards and accidents.

3) Plaintiffs simply proving that defective and unreasonably dangerous product elements existed at the time of shipment which contributed to the harm or injury.

Product Recall
When product dangers become evident, product recall has become a common mode of response by the manufacturer. Product recall should be avoided, however, due to its excessively costly nature.

Product-Safety Insurance
Insurance carriers for product risks increased their rates after a period of extensive product litigation. The carriers have now become more selective in underwriting products, requiring that product-safety standards be demonstrated for insurability. The Insurance Company of North America, for example, has outlined procedures for manufacturers to establish insurability as follow:

1) Study of the types of accidents, severity of injury, property damage, business interruptions and extra expenses likely that can be attributed to any possible risks in the use of the product.

2) Determination of measures that can be taken to reduce the risk and minimize the potential damage.

3) Determination of safety-development standards for the product.

4) Study of the human factors in the design of the product, including limitations of the users and their tendency to misuse the product or to use it in ways for which it was not intended.

Insurance engineers with the underwriter's staff assure that at least these minimal safety stardards are met when insurance coverage is to be provided.

Human-Error Risks in Product Usage

Human error can occur whenever human operators are involved. Human error, however, can generally be controlled when enough time is available, since the HO can catch the error before injury or damage occurs. For that reason, perhaps, human factors have typically been neglected in product safety considerations, since it has been believed that the HO's adaptability is sufficient to adjust for any design deficiency or product failure that might occur. Such an assumption is, of course, dangerous when significant risks are involved in the design configuration. These risks of human error can be significantly reduced by:

- Employing sound principles of control-display design and layout proven to be most efficient and error-free for human operation.
- Not assuming that safety problems are primarily due to equipment-failure risks. A more realistic outlook includes the possibility that human error can be a most significant consideration for not compromising product safety (Lowrance, 1976).

Product-Safety Analysis

Human factors analyses for safety programs should include at least the following:

1) First, a functional breakdown of the process, or just how the product is used.

2) A detailed account of how HOs interact with the product.

3) A determination of under what conditions the product is used.

4) An analysis of what operator actions should be when a dangerous condition occurs, e.g., disconnecting a power source in the case of fire.

5) A determination of the level of consumer information and knowledge required for safe product operation, e.g., it should not be assumed that everyone knows enough *not* to put their foot under a lawnmower housing.

6) An identification of all failures that can be expected in product operation and care, e.g., forgetting to put a cover back on a fill point.

7) An analysis of failure linkages and consequences in the possible chain of events that can lead to unsafe conditions, such as when a fuel cap is left off, resulting in fuel spillage on the hot engine, which results in a flame-up, which results in an explosion, etc.

173

Product Safety Design Considerations

In findings derived from the safety analysis, a number of design considerations should be made, including the following:

1) *Foolproofing.* Incorporate design features that lessen risks, e.g., covering all moving parts to prevent clothing from becoming entangled.

2) *Safety instructions.* Provide necessary instructions to assure safe behaviors on the part of the consumer. An analysis of the typical minimal experiences and educational background of the consumer population should permit a determination of reading level requirements. Instructional materials should be commensurate with the lowest comprehension level.

3) *Labeling.* Carefully selected words and codes should be used to describe potential dangers on placards, operating instructions, etc., in bold-letter warnings.

4) *Improvement feedback.* Correct any design features that are potentially hazardous. Specific failure data should be used to introduce design corrections and to prevent future repeated occurrences.

5) *Open disclosure.* Disclose any potentially hazardous conditions to make users fully aware of the risks.

6) *Legal protection.* Apply legal expediencies when any overtly hazardous practices are even remotely possible, i.e., state warnings in legally valid terms such as, "This lawnmower must never be lifted off the ground while operating it, or severe injury can result," etc.[1]

A Human Factors Program for Insurance Carriers

Insurance companies are sometimes criticized for being remiss in using substantive accident data. The primary function of insurance carriers, of course, is to reduce the burden of individual accidents on the insured party by spreading the costs over a large number of insured parties. However, the carriers can be seen to be in an excellent position to identify where the risks are from how accidents occur and to develop means to reduce the risk of such future occurrences.

Human factors studies, carried out in conjunction with insurance in-

---

1. Inherent safety design, of course, would make such a legal expediency unnecessary, e.g., the lawnmower automatically shuts down when it is lifted off the ground.

174

vestigations, might significantly help to reduce future accidents with the product or equipment. This, of course, would then reduce insurance rates as a secondary benefit (Burgess, 1979; Drury, 1983).

Summary

Comparatively recent legislation has made designers legally responsible for the safety of their products. The Consumer Product Safety Commission, created by the U.S. Congress in 1972, requires that product design be at a minimum of risk of injury to the consumer.

When product risks become evident, a common manufacturer's response has been one of recalling the product from the marketplace and from consumer usage. This solution is expensive and should be avoided. Rather, careful safety-design attention should be preferred prior to the release of the product.

Potential human error should be a major design consideration. A human factors analysis should be completed to minimize any product-safety risks, and should include a detailed consideration of how the user interacts with the product. Clearcut product-safety features and instructional provisions should be incorporated to reduce the risk of accident and injury.

A human factors analytical program is proposed for incorporation in insurance investigation procedures.

APPENDIX A
Acquiring a Statistical Sense

When interpreting descriptive human factors data, a designer may find that a statistical understanding is very helpful. In anthropometry, for example, the physical measurements obtained in a study are presumed to represent an entire population, such as adult males up to age 44, female adults over 75 years of age, etc. The available portion of these populations that can be measured is called a "sample." Of fundamental concern is whether or not the sample selected is statistically representative. If the anthropometric data are to be applied to the design of wheelchairs that are intended to accommodate an entire population of elderly persons, selected body sizes used for the design data must be completely representative or the design will simply not accommodate everyone.

In anthropometry, the sampling of body dimensions will—if a large enough sample is obtained—begin to approach a distribution of frequencies that form a bell-shaped curve (also called a "normal curve" or "Gaussian distribution").

The bell-shaped curve is derived from "random" sampling where no special effort has been made to select big individuals nor small ones; that is, they were, rather, picked on the basis of every size in the population at large having had an equal chance of being chosen.

The randomization of a sample can be illustrated by the "chance" factor that operates when a number of coins are flipped. To demonstrate this, have each member of the class throw out 10 pennies a given number of times. When the number of times (frequency) the total class has thrown out the ten pennies approaches 1,000 or 2,000 tosses, then a plot of these should look like the following:

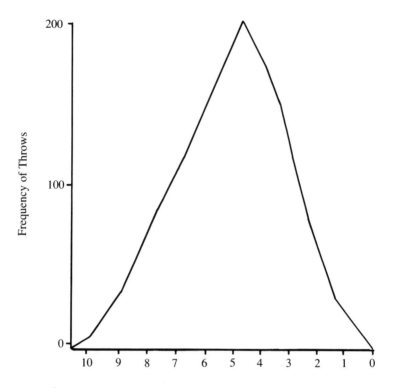

Number of Heads that Come Up Each Time the Coins Are Thrown Out

This exercise demonstrates the normal distribution or the frequency that occurs whenever selection is made on the basis of chance alone. Common statistical measurements used include the following:

$$\text{Average (also called the "mean")} = \frac{\Sigma\,(x)}{\text{Number of measures}}$$

Thus, to obtain the average or mean of all the measures, the sum (Σ) of the individual measures (x) is divided by the total number of individual measurements taken.

$$\text{Average deviation} = \frac{\Sigma\,(M - \text{-}x)}{\text{Number of measures}}$$

178

Here, the absolute sum of the amount that each individual measure *(x)* differs from the mean *(M)* is divided by the total number of measurements taken. Thus, for a mean height of 69 inches, a measurement obtained of 67 inches differs by 2 inches, a 73-inch measurement differs by 4 inches, and so on. These are summed and divided by the total number of such measurements.

$$\text{Standard deviation (SD)} = \sqrt{\frac{\Sigma(M-x)^2}{N}}$$

The SD is thus the square root of the sum (Σ) of each measure *(x)* as it differs from the mean *(m)*. It is then squared and the sum of these is divided by the total number of measurements taken.

The SD is the most important measure in sampling statistics since it describes chance variations within the normal curve. Any single score or measure *(x)* can then be related to the total distribution relative to where it falls above or below the mean. This can again be illustrated by the height measurement, where $M = 69$ inches, an individual measure of height is 72 inches, and the SD is 1.8 inches. Thus,

$$\frac{(M\text{-}x)}{SD} = \frac{69-72}{SD} = \frac{3}{1.8} = 1.6 \text{ SDs}$$

A normal curve percentage breakdown of SD units can be illustrated as follows:

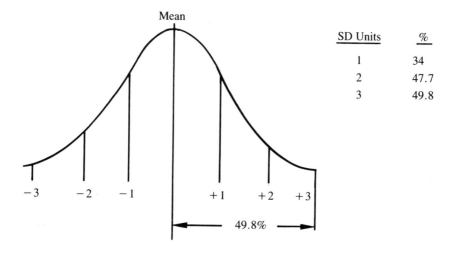

SD Units	%
1	34
2	47.7
3	49.8

So, 49.8 percent of all cases fall within 3 SDs above or below the mean, and 99.6 percent of all cases fall within ± 3 SDs. In the example of height given above, 1.6 SDs greater than the mean is where about 87 percent of all the cases occur, i.e., less than 72 inches tall. This is called the "87th percentile." Likewise, 66 inches in height, or 1.6 SDs below the mean, is where about 13 percent of all the cases are shorter. This called the "13th percentile."

The normal curve thus includes percentiles, based on standard deviation units above or below the means, as illustrated in Table A.1. Referring to the table, and using the case of height where the $SD = 1.8$ inches, the 99.9th percentile would be where only one person in a thousand would be taller. This would occur at 75.2 inches. The 0.1 percentile occurs at 62.7 inches, where only one person in a thousand would be shorter than this.

Table A.1. Percentiles with respect to units of the
standard deviation (SD).

Percentile		$(M \pm x)$
0.1		−3.49
5th		−2.44
10th		−1.90
15th		−1.54
20th		−1.25
25th		−1.00
30th		−0.78
35th		−0.57
40th		−0.38
45th		−0.19
50th	MEAN	0.00
55th		+0.19
60th		+0.38
65th		+0.57
70th		+0.78
75th		+1.00
80th		+1.25
85th		+1.54
90th		+1.90
95th		+2.44
99.9th		+3.49

Correlations

Statistical correlations are also based on this normal curve. In anthro-pometry, for example, height and weight would be significantly corre-lated in that the taller a person is, the more likely the weight will be greater. This is illustrated in the graph below.

These correlations can vary considerably for any given individual; a tall skinny person would not correlate on weight and height. When comparing body dimensions statistically, however, greater heights do average out to be correlated with greater body weights. A "perfect" cor-relation would be 1.00, whereas most dimensions correlate between around .60 and .90.

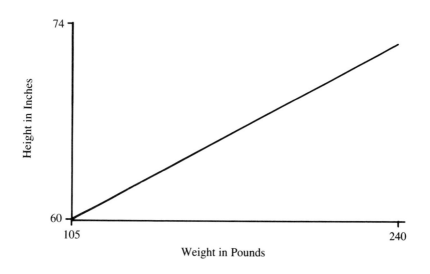

APPENDIX B
Some Industrial Pollutants

Industrial chemicals can be injurious to the HO by causing, for example, cellular and organ changes and by damaging DNA to produce birth defects. Numerous industrial processes generate pollutants. Whenever noxious elements are suspected, they should be investigated to determine their effects on health and safety. Following are common pollutants found in various industries:

Acrylonitrile ($CH_2 = CHCN$), a liquid pollutant in the plastic and synthetic-fiber industry. Toxic effects include nausea, diarrhea and excessive fatigue. Exposure should not exceed 2 parts per million.

Arsenic, a metalloid pollutant in such industries as coal-fueled power plants and copper, lead, and zinc smelters. Used in the production of glass, cloth, semiconductors, pesticides, and drugs. Symptoms following exposure include vomiting and dizziness, weakness, and a puffiness, reddening and thickening of the skin. Maximum limits are 10 micrograms per cubic liter of air over an eight-hour period.

Asbestos ($Mg_3Si_2O[OH]_4$) is sometimes called chrysotile or asbestos and is a mineral containing silicon, oxygen, hydrogen, sodium, magnesium and calcium. It is obtained through mining and is used for house products and insulation. Asbestos fibers are carcinogenic and can produce lung disease. Limits are 500,000 fibers per cubic meter of air over an eight-hour period.

Benzene (C_6H_6) is a by-product of petroleum and coke steel. Used in the manufacture of pesticides, inks, solvents, paints, plastics, gasoline, and pharmaceuticals. Exposure can result in fatigue and disorientation. Limits are one part per million.

Beryllium, a metal used in the manufacture of ceramics. Low dosage can cause bone and lung cancer and leukemia. Standards are two micrograms per cubic meter over an eight-hour period.

Cadmium, a soft white metal used in electroplating, paint manufacture, plastics, batteries, and metal-alloy products. Linked with prostate, lung and kidney cancer. Exposure limits should be less than 100 micrograms per cubic meter of air.

Carbon Tetrachloride (CCl_4) is a solvent used in pesticides, fire extinguishers and aerosols. Exposure produces nausea, vomiting and cramps. Limits are ten milligrams per cubic meter over an eight-hour period.

Chloroalkylethers, a family of chemicals used in plastics, rubber, and insecticides. May produce lung cancer. Exposure limits are 15 parts per million in the air over an eight-hour period.

Chromium, a heavy metal used in finishing metals and in dyes, electroplating, and leather tanning. It is carcinogenic and causes skin rash, nasal ulcers and bronchitis. Limits are one microgram per cubic meter of air.

Lead, a soft gray metal used in batteries, alloys, paint pigments, solder, gasoline additives, ceramic glazers, and metal cans. Exposure results in loss of appetite, listlessness, gastro-intestinal disturbances and convulsions. Limits are 0.1 milligrams per cubic meter of air.

Mercury, or quicksilver, is a silvery white metal fluid used in chlorine and caustic soda, lamp switches, batteries, thermometers, and paints. Exposure results in irritability, tremors and mouth disease. Limits are .05 milligrams per cubic meter of air.

Vinyl Chloride ($CH_2 = CHCl$) is a petrochemical used in such plastics as polyvinyl chloride and in the production of phonographs, electric wire, floor tiles, food wrappers, and automobile trim. It is a carcinogen and can cause skin rashes, finger numbness, and liver malfunction. Limits are one part per million in the air over an eight-hour period.

Glossary

AFTERIMAGE   Image that results after initial stimulation. In vision it is due to the chemical reaction of the eye and may be positive or negative.

ANTAGONIST MUSCLES   Attached to limbs in opposing action groups.

AUTONOMIC NERVOUS SYSTEM   Division of the nervous system supplying glands and viscera and regulating vital functions.

BALLISTIC MUSCLE ACTION   Rapidly thrown movements in contrast to slow, halting, constantly redirected tense movements.

COMPLEMENTARY COLORS   Two colors that, when mixed in light (not pigment), produce gray.

CONES   Color-sensitive structures in the retina.

CUTANEOUS   Pertaining to the skin.

DARK ADAPTATION   Increasing sensitivity of rods to light after some time in the dark.

EFFECTORS   Reacting muscles and glands.

EXTENSORS   Muscles which straighten out a limb.

FLEXORS   Muscles which bend a limb.

INNER EAR   Bone cavity inside the ear canal containing both auditory and balance sensory organs.

ISOMETRIC MUSCLE CONTRACTION   Measure of muscle tension made while muscle length is held constant.

ISOTONIC MUSCLE CONTRACTION   Measure of muscle contraction made while muscle shortens.

KINESTHETIC   Sense of motion (see PROPRIOCEPTORS).

NORMAL CURVE   Plot of frequencies based on probabilities that assumes a bell-shaped contour.

OLFACTORY   Sense of smell.

PARASYMPATHETIC   Cranial and sacral divisions of the autonomic nervous system.

PROPRIOCEPTORS   Sensory receptors in body-movement perception derived from stimulation of tiny spindles in the muscles and free nerve endings at the tendons and limb joints.

RECEPTORS   Sensory organs.

RODS   Organs in the retina that react only to light and are insensitive to colors.

SEMICIRCULAR CANALS   Three small channels in the inner ear, oriented about three planes of the body and containing fluid that stimulates tiny hair cells when the head moves.

STANDARD DEVIATION   An important measure in statistics that describes the way chance happenings or probabilities occur.

SUPPORT SUBSYSTEM   All the vital organs that carry nutrition and oxygen and dispose of waste in maintaining the sensory, information-processing and effector subsystems.

SYMPATHETIC   Central division of the autonomic nervous system.

SYNERGISTS   Cooperating muscles.

VESTIBULAR   Referring to the sensory organs used in head orientation in an upright position (see SEMICIRCULAR CANALS).

References

Acking, C. and R. Küller. "The Perception of an Interior as a Function of Its Color," *Ergonomics,* 1972, 15, 645-54.

Alderson, G. and H. Whiting. "Prediction of Linear Motion," *Human Factors,* 1974, 16, 495-502.

Allessi, S. and S. Trollip. *Computer-based Instruction. Methods and Development*, Englewood Cliffs, N.J.: Prentice Hall, 1985.

Baker, C. and W. Grether. "Visual Presentation of Information" in *Human Engineering Guide to Equipment Design* (C. Morgan, Ed.), New York: McGraw-Hill Book Company, 1963.

Bendix, T. "Seated Trunk Posture at Various Seat Inclinations, Seat Heights and Table Heights," *Human Factors,* 1984, 26, 695-703.

Bennett, C. and P. Rey. "What's So Hot About Red?," *Human Factors,* 1972, 14, 149-54.

Berry, P. "Effect of Colored Illumination Upon Perceived Temperature," *Journal of Applied Psychology,* 1961, 45, 248-50.

Birren, F. *Color Psychology and Color Therapy,* New Hyde Park, N.Y.: University Books, Inc., 1965.

Bromley, B. *The Psychology of Human Aging,* Baltimore, Md.: Penguin Books, 1966.

Bryson, A. and Y. Ho. *Applied Optimal Control,* New York: John Wiley and Sons, Inc., 1975.

Bullock, M. "The Determination of Functional Arm Reach Boundaries for Operating Manual Controls," *Ergonomics,* 1974, 17, 375-88.

Burgess, J. "Human Factors in the Insurance Industry," *Human Factors Society Bulletin,* April, 1979, 1-2.

Burgess, J. "The Use of Word Alarms in Industrial Emergencies," *Best's Safety Director,* 1983, 2, 1273-75.

Burgess, J. *Human Factors in Forms Design,* Chicago: Nelson Hall, Inc., 1984.

Burke, D. and C. Gibbs. "A Comparison of Free Moving and Pressure Levers in a Positional Control System," *Ergonomics,* 1965, 8, 23-29.

Busch, P. "A Review and Critical Evaluation of Consumer Product Safety Commission Marketing Management Implications," *Journal of Marketing,* October, 1976, 21-23.

Chapanis, A. "On the Allocation of Functions Between Men and Machines," *Occupational Psychology,* 1965, 39, 1-11.

Christ, R. "Review and Analysis of Color Coding Research for Visual Displays," *Human Factors,* 1975, 17, 542-70.

Clark, T. and E. Corlet. *The Ergonomics of Workspaces and Machines: A Design Manual,* London: Taylor and Francis, 1984.

Cochran, D. and M. Riley. "The Effects of Handle Shape and Size on Exerted Forces," *Human Factors,* 1986, 28, 253-265.

Cone, B., et al. "An Economic Analysis of Shovels and Draglines Used in U.S. Surface Coal Mines," Battelle Pacific Northwest Laboratories, June, 1976.

Cralley, L. and L. Cralley (Eds.). *Patty's Industrial Hygiene and Toxicology,* New York: John Wiley and Sons, Inc., 1985.

Cranston, H. "Magnetic Device Detects Bombs, Poisons, Cancer," *Moneysworth,* Summer, 1986, 2 ff.

Croney, J. *Anthropometry for Designers,* New York: Van Nostrand Reinhold, 1971.

Damos, D., et al. "Effects of Extended Practice on Dual-Task Tracking," *Human Factors,* 1981, 23, 627-31.

Department of Defense. "Military Handbook of Anthropometry of U.S. Military Personnel," DOD HDBK 743, Washington, D.C., 1980.

Department of Defense. "Military Standard 1472. Human Engineering Design Criteria for Military Systems, Equipment and Facilities," Washington, D.C., 1981.

Drury, C. "Human Factors in Consumer Product Accident Investigation," *Human Factors,* 1983, 25, 329-42.

Drury, C., et al. "A Survey of Industrial Box Handling," *Human Factors,* 1982, 24, 553-565.

Eastman-Kodak Human Factors. *Ergonomics Design for People at Work*, Belmont, Calif.: Wadsworth Publishing Company, 1983.

Edwards, E. and F. Lee. *Man and Computer in Process Control*, London: Institute of Chemical Engineering, 1972.

Ehrich, R. and R. Williges. *Advances in Human Factors/Ergonomics*, Amsterdam, Holland: Elsevier Publishing, 1986.

Emanuel, S., et al. "In Search of a Better Handle" in *Human Factors and Industrial Design in Consumer Products* (H. Poydar, Ed.), Medford, Mass.: Tufts University, Department of Engineering Design, 1980.

Eysenck, H. "A Critical and Experimental Study of Color Preferences," *American Journal of Psychology*, 1941, 54, 385-94.

Faulkner, T. and T. Murphy. "Lighting for Difficult Tasks," *Human Factors*, 1973, 15, 149-62.

Fozard, J. and S. Popkin. "Optimizing Adult Development. Ends and Means of An Applied Psychology of Aging," *American Psychologist*, 1978, 33, 975-89.

Goldman, A. and T. Slattery. *Maintainability: A Major Element of System Effectiveness*, New York: John Wiley and Sons, Inc., 1964.

Grieve, D., et al. *Techniques for the Analysis of Human Movement*, Princeton, N.J.: Princeton Book Company Publishers, 1975.

Hammond, R., et al. *Engineering Graphics Design, Analysis, Communication*, Huntington, N.Y.: Robert E. Krieger Publishing Co., 1979.

Hertzberg, H. "Engineering Anthropometry" in *Human Engineering Guide to Equipment Design* (H. Van Cott and R. Kinkade (Eds.), Washington, D.C.: U.S. Government Printing Office, 1972.

Hoffman, J. "No Constipation in Hunza," in *Hunza*, Hunza, Calif.: Professional Press Publishing Company, 1979, 190-200.

Hughes, P. and R. Neer. "Lighting for the Elderly. A Psychobiological Approach to Lighting," *Human Factors*, 1981, 23, 63-86.

Human Factors Bulletin. "The Bathroom. Room for Improvement," 1976, 19, April, 1-2.

Kira, A. *The Bathroom*, New York: Viking Press, 1976.

Kircher, J. "Product Liability of Design Engineers," *Professional Engineer*, January, 1970, 24-26.

Kroemer, K. and W. Marras. "Evaluation of Maximal and Submaximal Muscle Exertions," *Human Factors*, 1981, 23, 643-53.

Kryter, K. *The Effects of Noise on Man*, Orlando, Florida: Academic Press, 1985.

Kugath, D. "Designing Industrial Manipulators for the Consumer," *Proceedings of the 17th Annual Meeting of the Human Factors Society*, Santa Monica, Calif., 1973, 315-21.

Kvalseth, T. *Ergonomics of Workstation Design*, Stoneham, Mass.: Butterworth Publishing Company, 1983.

Leibowitz, W. and D. Owens. "Nighttime Driving Accidents and Selective Visual Degradation," *Science*, 1977, 197, 422-23.

Leibowitz, H. and D. Owens. "We Drive by Night," *Psychology Today*, January, 1986, 54-58.

Lowrance, W. *Of Acceptable Risk. Science and the Determination of Safety*, Los Altos, Calif.: William Kaufmann, Inc., 1976.

Luce, P., et al. "Capacity Demands in Short-Term Memory for Synthetic and Natural Speech," *Human Factors*, 1983, 25, 17-32.

Martin, W. *Basic Body Dimensions of School Age Children*, Washington, D.C.: U.S. Department of Health, Education and Welfare, 1953.

McClelland, I. "The Ergonomics of Toilet Seats": *Human Factors*, 1982, 24, 713-725.

McCrobie, D. "Application of MacProject to the Task Analysis Process," *Human Factors Society Bulletin*, 1986, 29, 5.

Miller, R. "A Method for Man-Machine Task Analysis," Wright-Patterson Air Force Base, Ohio: WADC TR 53-137, 1953.

Miller, G. *Language and Communication*, New York: McGraw-Hill Book Company, 1963.

Mishkin, M., et al. "Object Vision and Spatial Vision: Two Cortical Pathways," *Neurosciences*, October, 1983, 6, 414-17.

Morgan, C. (Ed.). *Human Engineering Guide to Equipment Design*, New York: McGraw-Hill Book Company, 1963.

Mourant, R., et al. "Human Factors Requirements for Fingertip Reach Controls," Washington, D.C.: U.S. Department of Transportation, 1977.

National Bureau of Standards. "Body Measurements for the Sizing of Apparel for Infants, Babies, Toddlers and Children," *Common Standards*, CS151-50 June, 1953.

Nickerson, R. "Human Factors and the Handicapped," *Human Factors*, 1978, 20, 259-72.

O'Hare, J. "Human Factoring the Programmer's Task," *Human Factors Society Bulletin*, 1981, 25, 5.

Parsons, S., et al. "Human Factors Design Practices for Nuclear Power Plant Control Rooms," *Proceedings of the 22nd Annual Meeting of the Human Factors Society*, Santa Monica, Calif.: 1978.

190

Pisoni, D. and E. Koen. "Intelligibility of Natural and Synthetic Speech in Several Different Signal-to-Noise Ratios," *Journal of the Acoustical Society of America,* 1982, 71, UU1.

Price, H. "The Allocation of Functions in a System," *Human Factors,* 1985, 27, 33-46.

Ranney, T. and V. Gawron. "The Effect of Pavement Edgelines on Performance in a Driving Simulator under Sober and Alcohol-Dosed Conditions," *Human Factors,* 1986, 28, 511-526.

Roebuck, J., et al. *Engineering Anthropometry Methods,* New York: John Wiley and Sons, Inc., 1975.

Roscoe, S. "Designed for Disaster," *Human Factors Bulletin,* 1986, 29, 1-2.

Rosenthal, M. "Application of Human Engineering Principles and Techniques in the Design of Electronic Production Equipment," *Human Factors,* 1973, 15, 137-148.

Rozier, C. "Three dimensional Work Space of the Amputee," *Human Factors,* 1977, 19, 525-533.

Scheibe, K., et al. "Color Association Values and Response Interference on Variants of the Stroop Test," *Acta Psychologica,* 1967, 28, 286-95.

Schonfield, D. "Translations in Gerontology—From Lab to Life," *American Psychologist,* 1974, 29, 796-801.

Sheridan, T. and R. Mann. "Design of Control Devices for People with Severe Motor Impairment," *Human Factors,* 1978, 20, 321-38.

Sivak, M., et al. "Driver Responses to High Mounted Brake Lights in Actual Traffic," *Human Factors,* 1981, 23, 231-36.

Slowiaczek, L. and H. Nusbaum. "Effects of Speech and Pitch Contour on the Perception of Synthetic Speech," *Human Factors,* 1985, 27, 701-12.

Stoudt, H., et al. "Weight, Height and Selected Body Dimensions of Adults," *Vital and Health Statistics,* Series 11, Number 8, Rockville, Md.: Public Health Service, 1965.

Sundstrom, E. and M. Graehl. *Workplaces. Psychology of the Physical Environment in Offices and Factories,* New York: Cambridge University Press, 1986.

Tichauer, E. *Biomechanical Basis of Ergonomics,* New York: John Wiley and Sons, Inc., 1978.

VanCott, H. and R. Kinkade (Eds.). *Human Engineering Guide to Equipment Design,* Washington, D.C.: U.S. Government Printing Office, 1972.

Wade, M. and M. Gold. "Removing Some of the Limitations of Mentally Retarded Workers by Improving Job Design," *Human Factors,* 1978, 20, 339-48.

Warren, R. "Norms of Restricted Color Associations," *Bulletin of the Psychonomic Society,* 1974, 4, 37-38.

Weitzman, D., et al. "Proficiency, Maintenance and Assessment on an Instrument Flight Simulator," *Human Factors,* 1979, 21, 701-10.

Welford, A. "Signal Noise, Performance and Age," *Human Factors,* 1981, 23, 97-110.

White, R. "The Anthropometry of United States Men and Women: 1946-77," *Human Factors,* 1979, 21, 473-82.

Wierwille, W. and S. Connor. "Evaluation of 20 Workload Measurements Using a Psychomotor Task in a Moving Base Simulator," *Human Factors,* 1983, 25, 1-16.

Woodson, W. and D. Conover. *Human Engineering Guide for Equipment Designers,* Berkeley, Calif.: University of California Press, 1964.

Woodson, W. *Human Factors Design Handbook,* New York: McGraw-Hill Book Company, 1981.

Ziegler, P. "Single and Dual Axis Tracking as a Function of System Dynamics," *Human Factors,* 1968, 10, 273-76.

Additional Source Material

Anthropometry

Cochran, D., and M. Riley. "The Effects of Handle Shape and Size on Exerted Forces," *Human Factors*, 1986, 28, 253-265.

Croney, J. *Anthropometry for Designers*, New York: Van Nostrand Reinhold, 1971.

Damon, A., et al. *The Human Body in Equipment Design*, Cambridge, Mass.: Harvard University Press, 1966.

Dreyfuss, H. *The Measure of Man*, New York: Whitney Library of Design, 1967.

Differient, N., et al. *Humanscale*, 1 through 9, Cambridge, Mass.: MIT Press, 1974.

Easterby, R., et al., (Eds.). *Anthropometry and Biomechanics. Theory and Application*, New York: Plenum Publishing Company, 1981.

Kleeman, W. "A Different Way to Use Anthropometric Data as a Tool for Computer Terminal Workstation Design," *Human Factors Society Bulletin*, 1987, 30, 4-6.

Martin, W. *Basic Body Dimensions of School Age Children*, Washington, D.C.: U.S. Department of Health, Education and Welfare, 1953.

National Aeronautics and Space Administration. *Anthropometric Source Book*, Volumes 1, 2 and 3 (NASA Publication Number 1024), Washington, D.C.: U.S. Government Printing Office, 1979.

Panero, J. and M. Zelnick. *Human Dimension and Interior Space*, New York: Whitney Library of Design, 1979.

Pheasant, S. *Body Space Anthropometry, Ergonomics an Design,* Philadelphia, Penna.: Taylor and Francis, 1984.

Roebuck, J., et al. *Engineering Anthropometric Methods,* New York: John Wiley and Sons, Inc., 1975.

Stoudt, H. "The Anthropometry of the Elderly," *Human Factors,* 1981, 23, 29-38.

Computer Design

Alessi, S. and S. Trollip. *Computer-based Instruction. Methods and Development* Englewood Cliffs, N.J.: Prentice Hall, 1985.

Brown, C. *Human-Computer Interface Design Guidelines,* Norwood, N.J.: Ablex Publishing Corporation, 1986.

deHaan, H. (U.S. Army Research Institute, Springfield, Virginia). "Vocal Control of Computer Generated Map Displays," American Psychological Association 94th Convention, Washington, D.C., 1986

Ehrich, R. and R. Williges (Eds.). *Human-Computer Dialogue Design,* New York: Elsevier Science Publishers, 1986.

Guindon, R. (Ed.). *Cognitive Science and Its Applications for Human-Computer Interaction,* Hillsdale, N.J.: Lawrence Erlbaum Associates, Inc., 1987.

National Research Council. *Video Displays, Work and Vision,* Washington, D.C.: National Academy Press, 1983.

Norman, D. and S. Draper (Eds.). *User Centered System Design. New Perspectives on Human-Computer Interaction,* Hillsdale, N.J.: Lawrence Erlbaum Associates, Inc., 1986.

Savendy, G. and A. Majchrzak. *Human Aspects of Computer Aided Design,* Philadelphia, Penna.: Taylor and Francis, 1987.

Simpson, C., et al. "System Design for Speech Recognition and Generation," *Human Factors,* 1985, 27, 115-42.

Thomas, J. and M. Schneider. *Human Factors in Computer Systems,* Norwood, N.J.: Ablex Publishing Corporation, 1984.

Design Applications

Babbs, F. "A Design Layout Method for Relating Seating to the Occupant and Vehicle," *Ergonomics,* 1979, 22, 227-34.

Bales, J. "Human Factors Studies Help to Make Skies Safer," *American Psychological Association Monitor,* 1986, 17, 10.

Brooks, R. "An Investigation in Aspects of Bus Design and Passenger Requirements" *Ergonomics,* 1979, 22, 175-78.

Burgess, J. *Human Factors in Built Environments*, Newtonville, Mass.: Environmental Design and Research Center, 1981.

Burgess, J. *Human Factors in Forms Design*, Chicago: Nelson Hall, Inc., 1984.

Christie, B. *Human Factors of Information Technology in the Office*, New York: John Wiley and Sons, Inc., 1985.

Connors, M., et al. "Psychology and the Resurgent Space Program," *American Psychologist*, 1986, 41, 906-13.

Crittendent, L. "Some Thoughts on the Use of Synthetic Speech for Cockpit Warnings," *Human Factors Society Bulletin*, 1986, 29, 3-4.

Gay, K. *Ergonomics: Making Products and Places Fit People*, Hillside, N.J.: Enslow Press, 1986.

Gordon, R. *Microelectronics in Transition. Industrial Transformation and Social Change*, Norwood, N.J.: Ablex Publishing Corporation, 1986.

Grandjean, E. *Ergonomics of the Home*, Philadelphia, Penna.: Taylor and Francis, 1978.

Grandjean, E. (Ed.). *Ergonomics and Health in Modern Offices*, New York: Taylor and Francis, 1984.

Institution of Chemical Engineers. *Ergonomics Problems in Process Operations*, Elmsford, N.Y.: Pergamon Press, 1984.

Jahns, D. "Getting It All Together: A Case Study in Human Factors-Industrial Design Engineering Technology," *Human Factors Society Bulletin*, 1985, 28, 1-3.

Kira, A. *The Bathroom*, New York: Bantam Books, 1966.

Kvalseth, T. *Ergonomics of Workstation Design*, Stoneham, Mass.: Butterworth Publishing Company, 1983.

Lupton, T. (Ed.). *Human Factors in Manufacturing. Proceedings of the 1st International Human Conference*, New York: Elsevier Science Publishing Company, Inc., 1984.

Maule, H. and J. Weiner. *Design for Work and Use. Case Studies in Ergonomics Practice*, Philadelphia, Penna.: Francis and Taylor, 1981.

Morrison, R., et al. "Movement Time and Brake Pedal Placement," *Human Factors*, 1986, 28, 241-46.

Otway, H. and M. Peltu (Eds.). *New Office Technology. Human and Organizational Aspects*, Norwood, N.J.: Ablex Publishing Corporation, 1983.

Pile, J. *Modern Furniture*, New York: John Wiley and Sons, Inc., 1979.

Shneiderman, B. *Designing the User Interface: Strategies for Effective Human-Computer Interaction*, Reading, Mass.: Addison-Wesley Publishing Company, 1986.

Sundstrom, Eric, University of Tennessee at Knoxville. "Physical Environments in Offices and Factories. Current Research and Future Priorities," paper presented at the 1985 American Psychological Association Convention in Los Angeles.

Tichauer, E. *The Biomechanical Basis of Ergonomics*, New York: John Wiley and Sons, Inc., 1978.

Design Guidelines

Applied Ergonomics Handbook. Surrey, England: IPC Science and Technology Press Ltd., 1977.

Ayoub, M. "Workplace Design and Posture," *Human Factors*, 1973, 15, 265-68.

Christie, B. (Ed.). *Human Factors of the User-System Interface*, New York: Elsevier Science Publishing Co., Inc., 1985.

Keppel, G. *Design and Analysis. A Researcher's Handbook*, Englewood Cliffs, N.J.: Prentice Hall, 1973.

Kinkade, R. and H. Van Cott. *Human Engineering Guide to Equipment Design*, Washington, D.C.: U.S. Government Printing Office, 1972.

Lang, B., et al. *Designing for Human Behavior*, Stroudsburg, Penna.: Hutchinson-Ross Publishing Company, 1974.

Roberts, S. *Industrial Design With Microcomputers*, Englewood Cliffs, N.J.: Prentice Hall, Inc., 1982.

Sommer, R. and B. Sommer. *A Practical Guide to Behavioral Research*, New York: Oxford University Press, 1980.

Tullis, T. and B. Bied. *Space Station Functional Relationships Analysis*, Huntington Beach, Calif.: McDonnell Douglas Corp., 1986.

Wise, J. *The Quantitative Modeling of Human Spatial Habitability*, Moffett Field, Calif.: NASA-Ames Research Center, 1985.

Woodson, W. *Human Factors Design Handbook*, New York: McGraw-Hill Book Company, 1981.

Designing for the Disabled and Impaired

Bazar, A. "Grip Strength of Cerebral Palsied," *Human Factors*, 1978, 20, 741-44.

Charneso, N. (Ed.). *Aging and Human Performance*, New York: John Wiley and Sons, Inc., 1985.

Harkness, S. and J. Groom. *Building Without Barriers for the Disabled*, New York: Whitney Library of Design, 1976.

Koncelik, J. *Aging and the Product Environment,* Stroudsburg, Penna.: Hutchinson-Ross Publishing Company, 1982.

Koncelik, J. *Designing the Open Nursing Home,* Stroudsburg, Penna.: Hutchinson-Ross Publishing Company, 1976.

Rozier, C. "Three-Dimensional Work Space for the Amputee," *Human Factors,* 1977, 19, 425-34.

Evaluation Techniques

Burden, E. *Design Simulation: Uses of Photographic and Electronic Media in Design and Presentation,* New York: John Wiley and Sons, Inc., 1984.

Childs, J. "Time and Error Measures of Human Performance," *Human Factors,* 1980, 22, 113-17.

Cox, C. *A Handbook of Introductory Statistical Methods,* Somerset, New Jersey: John Wiley & Sons, Inc., 1987.

Easterby, R. and H. Zwaga. "Evaluation of Public Information Symbols," University of Aston, Birmingham, England. Applied Psychology Department, International Organization for Standardization Tests. Report No. 60, 1976.

Frey, A. and E. Eickert. "An Evaluation of Holograms as Training and Job Aids," *Human Factors,* 1978, 20, 661-70.

Khalil, T. "An Electromyographic Methodology for the Evaluation of Industrial Design," *Human Factors,* 1973, 15, 257-64.

Meister, D. and G. Rabideau, *Human Factors Evaluation in System Development* New York: John Wiley & Sons, Inc., 1965.

National Research Council. *Human Factors Aspects of Simulation,* Washington, D.C.: National Academy Press, 1985.

Roberts, N., et al. *Fault Tree Handbook,* NUREG-0492, Washington, D.C.: Nuclear Regulatory Commission, 1980.

Siegel, A. and J. Wolf. *Man-Machine Simulation Models,* New York: John Wiley & Sons, Inc., 1969.

Whealer, J. "Evaluation of an Experimental Central Warning System on a Synthesized Voice Component," *Aviation Space and Environmental Medicine,* 1983, 54, 517-23.

Wilson, J. and N. Corlelett. *Ergonomics of Working Postures,* Philadelphia, Penna.: Taylor and Francis, 1986.

General Human Factors/Ergonomics

Bailey, R. *Human Performance Engineering: Guidelines for Design,* Englewood Cliffs, N.J.: Prentice Hall, 1982.

Bennett, E., et al. *Human Factors in Technology,* New York: McGraw-Hill Book Company, 1963.

Burgess, J. *Designing for Humans: The Human Factor in Engineering* Princeton, N.J.: Petrocelli Books, 1986.

Chapanis, A. *Man-Machine Engineering,* Belmont, Calif.: Wadsworth Publishing Company, 1965.

Corlett, E., et al. (Eds.) *Methods in Applied Ergonomics,* New York: Taylor and Francis, 1987.

Clark, T. and E. Corlett. *The Ergonomics of Workspaces and Machines,* Philadelphia, Penna.: Taylor and Francis, 1984.

deMontmollin, M. and L. Bainbridge. "Ergonomics or Human Factors," *Human Factors Society Bulletin,* 1985, 28, 1-3.

Department of Defense. "Human Engineering Design Data Pocket Digest of MIL STD 1472," 1984.

Dreyfus, N. and S. Dreyfus. *Mind Over Machine,* New York: Free Press, 1983.

Dreyfuss, H. *Designing for People,* New York: Viking/Compass Press, 1955.

Dreyfuss, H. *The Measure of Man: Human Factors in Design,* New York: Whitney Library of Design, 1967.

Gay, K. *Ergonomics: Making Products and Places Fit People,* Hillside, N.J.: Enslow Publishing Company, Inc., 1985.

Grandjean, E. *Fitting the Task to the Man,* New York: International Publications Service, 1981.

Huchingson, R. *New Horizons for Human Factors in Design,* New York: McGraw-Hill Book Company, 1981.

Kantowitz, B. and R. Sorkin. *Human Factors,* New York: John Wiley and Sons, Inc., 1983.

McCormick, E. and M. Sanders. *Human Factors in Engineering and Design,* New York: McGraw-Hill Book Company, 1982.

Megaw, E. (Ed.). *Contemporary Ergonomics 1987—Ergonomics Working for Society,* New York: Taylor and Francis, 1987.

Murrell, K. *Ergonomics: Man and His Working Environment,* London: Chapman and Hall, Publishers, 1969.

Perrow, C. "The Organizational Context of Human Factors Engineering," *Administrative Science Quarterly,* 1983, 28, 521-24.

Salvendy, G. (Ed.). *Handbook of Human Factors,* New York: John Wiley and Sons, Inc., 1986.

Singleton, W. *Introduction to Ergonomics,* Geneva, Switzerland: World Health Organization.

Tichauer, E. *The Biomechanical Basis of Ergonomics*, New York: John Wiley and Sons, Inc., 1978.

Wickens, C. *Engineering Psychology and Human Performance*, Columbus, Ohio: Charles Merrill Publishing Company, 1984.

Woodson, W. *Human Factors Design Handbook*, New York: McGraw-Hill Book Company, 1981.

Graphics Design

Cahill, M. "Interpretability of Graphic Symbols as a Function of Context and Experience Factors," *Journal of Applied Psychology*, 1975, 60, 376-80.

Dick, W. and L. Carey. *The Systematic Design of Instructions*, Glenview, Illinois: Scott, Foresman and Company, 1978.

Easterby, R. (Ed.). *Information Design. The Design and Evaluation of Signs and Printed Material*, New York: John Wiley & Sons, Inc., 1984.

Dreyfuss, H. *Symbol Sourcebook*, New York: McGraw-Hill Book Company, 1972.

Howell, W. and A. Fuchs. "Population Stereotype in Code Design," *Organizational Behavior and Human Performance*, 1968, 3, 310-39.

Marshall, G. and C. Cofer. "Associative Indices as Measures of Word Relatedness: A Summary of Ten Methods," *Journal of Verbal Behavior*, 1963, 1, 408-21.

Modley, R. *Handbook of Pictorial Symbols*, New York: Dover Press, 1976.

Murch, G. *The Effective Use of Color: Cognitive Principles*, Tektronix, Inc., 1984.

Posner, M., et al. "Retention of Visual and Name Codes of Single Letters," *Journal of Experimental Psychology Monographs*, 1969, 79, 1-16.

Spence, G. "Human Factors in Interactive Graphics," *Computer Aided Design*, 1976, 8, 49-53.

Zwaga, H. *Research on Graphic Symbols*, Netherlands: University of Utrecht, Psychological Laboratory Report #74-4, 1974.

Handtool Design

Cochran, D. and M. Riley. "The Effects of Handle Shape and Size on Exerted Forces," *Human Factors*, 1986, 28, 253-66.

Drury, C. "Handles for Manual Material Handling," *Applied Ergonomics*, 1980, 11, 35-42.

Knowleton, R. and J. Gilbert. "Ulnar Deviation and Short Term Strength Reduction as Affected by a Curved Handle, Ripping Hammer and Conventional Claw Hammer," *Ergonomics*, 1983, 26, 173-79.

Konz, S. "Bent Hammer Handles," *Human Factors,* 1986, 28, 317-24.

Kroemer, K. "Coupling the Hand With the Handle: An Improved Notation of Touch, Grip and Grasp," *Human Factors,* 1986, 28, 337-40.

Mital, A. (Ed.). "Special Issue: Hand Tools," *Human Factors,* 1986, 28, 251-373.

Mital, A. and N. Sanghavi. "Comparison of Maximum Volitional Torque Exertion Capabilities of Males and Females Using Common Hand Tools," *Human Factors,* 1986, 28, 283-294.

Tichauer, E. and H. Gage. "Ergonomic Principles Basic to Hand Tool Design," *American Industrial Hygiene Association Journal,* 1977, 38, 622-34.

Human Performance

Beyth-Marom, R., et al. *An Elementary Approach to Thinking Under Uncertainty,* Hillsdale, N.J.: Lawrence Erlbaum Associates, Inc., 1985.

Birren, J. and K. Schaie (Eds.). *Handbook of the Psychology of Aging,* New York: Van Nostrand Reinhold, 1985.

Boff, R., et al. (Eds.). *Handbook of Perception and Human Performance,* Volumes 1 and 2, New York: John Wiley and Sons, Inc., 1986.

Chaffin, D. and G. Anderson. *Occupational Biomechanics,* New York: John Wiley and Sons, Inc., 1984.

Davies, R. and R. Parasuraman (Eds.). *The Psychology of Vigilance,* New York: Academic Press, 1982.

Dhillon, B. *Human Reliability with Human Factors,* Elmsford, N.Y.: Pergamon Press, 1986.

Grinker, R. and J. Spiegel. *Men Under Stress,* New York: McGraw-Hill Book Company, 1963.

Hockey, R. *Stress and Fatigue in Human Performance,* New York: John Wiley and Sons, Inc., 1983.

Hull, D., et al. "Relaxed + G Tolerance in Healthy Men. Effects of Age," *Journal of Applied Physiology,* 1978, 45, 426-29.

Kra, S. *Aging Myths: Reversible Causes of Mind and Memory,* New York: McGraw-Hill Book Company, 1986.

Moray, E. (Ed.). *Mental Workload. Its Theory and Measurement* New York: Plenum Publishing Co., 1981.

Posner, M. and O. Marin (Eds.). *Mechanisms of Attention. Attention and Performance XI,* Hillsdale, N.J.: Lawrence Erlbaum Associates, Inc., 1985.

Rasmussen, J. *Information Processing and Human-Machine Interaction. An Approach to Cognitive Engineering,* New York: Elsevier Science Publishing Co., Inc. 1986.

Singleton, W. and J. Hovden (Eds.). *Risk and Decisions,* Somerset, New Jersey: John Wiley & Sons, Inc., 1987.

Slater, K. *Human Comfort,* Springfield, Illinois: Charles C. Thomas, Publisher, 1985.

Trotter, R. "The Mystery of Mastery," *Psychology Today,* July, 1986, 32-38.

Warm, J. (Ed.). *Sustained Attention in Human Performance,* New York: John Wiley & Sons, Inc., 1984.

Warm, J. and W. Dember. "Awake at the Switch," *Psychology Today,* April, 1986, 46-53.

Welford, A. (Ed.). *Men Under Stress,* New York: Halsted Press, 1974.

Maintainability

Goldman, A. and T. Slattery. *Maintainability,* New York: John Wiley & Sons, Inc., 1964.

Perceptual Factors

Baker, C. and W. Grether. "Visual Presentation of Information" in *Human Engineering Guide to Equipment Design* (C. Morgan, Ed.), New York: McGraw-Hill Book Company, 1963.

Boff, K., et al. (Eds.). *Handbook of Perception and Human Performance,* Somerset, New Jersey: John Wiley & Sons, Inc., 1986.

Boyce, P. *Human Factors in Lighting,* Riverside, N.J.: Macmillan Publishing Company, Inc., 1981.

Burgess, J. "Improving the Effective Intelligibility of Emergency Warning Signals," *National Safety News,* 1981, 123, 33-34.

Burgess, J. "The Use of Word Alarms in Industrial Emergencies," Oldwick, N.J.: *Best's Safety Directory,* 1983.

Christ, R. "Review and Analysis of Color Coding Research for Visual Displays," *Human Factors,* 1975, 17, 542-70.

Dember, W. and J. Warm. *The Psychology of Perception,* New York: Holt, Rinehart and Winston, 1979.

Easterby, R. and H. Zwaga (Eds.). *Visual Presentation of Information. The Design and Evaluation of Signs and Printed Material,* New York: John Wiley and Sons, Inc., 1983.

Frier, J. and M. Gazley. *Industrial Lighting Systems,* New York: McGraw-Hill Book Company, 1981.

Gregory, R. *Eye and Brain. The Psychology of Seeing,* London: World University Press Library, 1966.

201

IBM Corporation. "Human Factors of Workstations with Visual Displays," San Jose, Calif.: IBM Corporation, 1978.

Pinker, S. (Ed.) *Visual Cognition,* Cambridge, Mass.: MIT Press, 1984.

Shontz, W., et al. "Color Coding for Information Location," *Human Factors,* 1971, 13, 237-46.

Sundgaard, E. *Human Form Visual Reference,* New York: Van Nostrand Reinhold, 1980.

Weintraub, D. and E. Walker. *Perception,* Belmont, Calif.: Wadsworth Publishing Company, 1966.

Physiological Factors in Design

Davis, H., et al. "Work Physiology," *Human Factors,* 1969, 11, 157-66.

Dawson, H. and M. Segal. *Introduction to Physiology,* New York: Grune and Stratton, 1975.

Hockey, R. *Stress and Fatigue in Human Performance,* New York: John Wiley & Sons, Inc., 1983.

Mackie, R. (Ed.). *Vigilance: Theory, Operational Performance and Physiological Correlates,* New York: Plenum Publishers, 1977.

Product Safety

Bass,L. *Products Liability: Design and Manufacturing Defects,* Colorado Springs, Colorado: Shepard's/McGraw-Hill, Inc., 1986.

Kolb, J. and S. Ross. *Product Safety and Liability Desk Reference,* New York: McGraw-Hill Book Company, 1981.

Sherman, P. *Products Liability,* Colorado Springs, Colorado: Shepard's/McGraw-Hill, Inc., 1985.

Robotics

Butera, F. (Ed.). *Automation and Work Design,* New York: Elsevier Science Publishing Company, Inc., 1984.

Noro, K. (Ed.). *Occupational Health and Safety in Automation and Robotics,* Philadelphia, Penna.: Taylor and Francis, 1986.

Schwartz, J. and M. Sharir (Eds.). *Planning, Geometry and Complexity of Robot Motion,* Norwood, N.J.: Ablex Publishing Corporation, 1986.

Simons, G. *Is Man a Robot?,* Somerset, New Jersey: John Wiley & Sons, Inc., 1986.

Statistics

Hopkins, K. and G. Glass. *Basic Statistics for the Behavioral Sciences,* Englewood Cliffs, N.J.: Prentice Hall, Inc., 1978.

Klugh, H. *Statistics: Essentials for Research,* Hillsdale, N.J.: Lawrence Erlbaum Associates, Inc., 1986.

Stereotypes

Courtney, A. "Chinese Population Stereotypes: Color Association," *Human Factors,* 1986, 28, 97-99.

Systems

Burgess, J. "Ego Involvement in the System Design Process," *Human Factors,* 1970, 12, 7-12.

Christie, B. (Ed.). *Human Factors on the User-System Interface: A Report on Esprit Preparatory Study,* New York: Elsevier North Holland Publishing Co., 1985.

Fitts, P. "Functions of Man in Complete Systems," *Aerospace Engineering,* 1962, 21, 34-39.

Hamstra, N. and V. Ellingstad. *Human Behavior: A Systems Approach,* Monterey, Calif.: Brooks Cole Publishing Company, 1972.

Kruglanski, A. "Freeze Think and the Challenger," *Psychology Today,* August, 1986.

Meister, D. *Behavioral Foundations of Systems Development,* New York: John Wiley & Sons, Inc., 1976.

Michle, D. *On Machine Intelligence,* Somerset, New Jersey: John Wiley & Sons, Inc., 1986.

Miller, R. "A Method for Man-Machine Task Analysis": Wright-Patterson Air Force Base, Ohio: WADC TR 53-137, 1953.

Moraal, J. and K. Kraus (Eds.). *Manned Systems Design Methods, Equipment and Applications,* New York: Plenum Publishing Corporation, 1985.

Oborne, D. and M. Gruneberg (Eds.). *The Physical Environment at Work,* New York: John Wiley and Sons, Inc., 1983.

Price, H. "The Allocation of Functions in Systems," *Human Factors,* 1985, 27, 47-60.

Pulliam, R. and H. Price. *Automation and the Allocation of Functions Between Human and Automatic Control: General Method,* Macaulay-Brown, Inc., 1985.

Roszak, T. *The Cult of Information. The Folklore of Computers and the True Art of Thinking,* New York: Pantheon Books, 1986.

Rouse, W. *Systems Engineering Model of Human-Machine Interaction,* New York: Elsevier North Holland Publishing Company, 1985.

Rouse, W. and K. Boff (Eds.). *System Design: Human and Technological Factors in the Design of Complex Systems,* New York: Elsevier Science Publishing Co., Inc., 1987.

Spettell, C. and R. Liebert. "Training for Safety in Automated Person-Machine Systems," *American Psychologist,* 1986, 41, 545-50.

Journals

Applied Ergonomics

Behavioral and Information Technology

Computer Speech and Language

Computers in Industry

Data and Knowledge Engineering

Decision Support Systems

Engineering Psychology

Ergonomics

Ergonomics Abstracts

Ergonomics Newsletter

Human-Computer Interaction

Human Performance

Human Factors and *Human Factors Society Bulletin*

Journal of Applied Psychology

International Journal of Man-Machine Studies

International Reviews of Ergonomics

Proceedings of the Human Factors Society Annual Meetings

Work and Stress

Contacts for Human Factors Workshops, Short Courses and Seminars

Society for Information Display
201 Varick Street, Room 1140
New York, N.Y. 10014

University of California at Los Angeles
UCLA Extension
Department of Engineering and Science
10995 LeConte Avenue
Los Angeles, Calif. 90024

The University of Maryland University College
Professional and Career Development Program
University Blvd. at Adelphi Rd.
College Park, Md. 20742

Engineering Summer Conferences
400 Chrysler Center, North Campus
The University of Michigan
Ann Arbor, Mich. 48109

The Continuing Education Institute
21250 Califa Street, Suite 102
Woodland Hills, Calif. 91367

Index

A

acceleration, 46, 91
accessibility, maintenance and, 151
acrylonitrile, 183
activity analysis (see task analysis)
adaptive training, 27, 28
additive system units, product improvement and, 107
adjunctive motion senses, 12
afterimages, 6
aged and impaired, design limitations for, 48-49
airport antihijacking security as complex systems, 140
aldehydes, 46
ambient pressures, 90
ammonia, 46
antagonistic movement, 9
anthropometry, 32-37
  age vs. height, weight, etc., 33
  common measurements in, 34
  design strategies using, 36-37
  designing for extremes, 37
  designing for the median, 36
  full-range strategies, 37
  sampling for, 177
  taking measurements for, 49-50
asbestos, 183
auditory displays, 77
  synthetic speech in, 78
automobile as complex systems, 139
autonomic nervous system, 13
average deviation, 178
averages, 178

B

balance, 4, 5, 12
ballistic movement, 11
bell-shaped curves, 177
benzene, 183
beryllium, 184
binocular vision, 7

biomechanical limitations, 38-41
black outs, 46
bodies (see human form and function)
boredom, 31
brain (see nervous system and brain)
brain dysrhythmia, 44
buzz vision, 44

C

cadmium, 184
capabilities, 17-28, 120
carbon monoxide, 45
carbon tetrachloride, 184
children and adolescents, design limitations for, 47-48
chloroalkylethers, 184
choroid, 5
chromium, 184
circulatory and digestive system, 13-14
color blindness, 7
color sensitivity, 7
communications, 119
complex machines, 53, 56
complex systems, 53, 56
  activity and task analysis, 146
  airport antihijacking security, 140
  allocating system functions in, 144
  analytical techniques for, 141
  automobile as, 139
  computer system parameter control, 140
  developing system mission profile for, 143
  development stages in, 138
  environmental stress factors and, 148
  functions analysis for, 144
  human factors applications in design of, 137-154
  information requirements analysis for, 146
  interface design, 147
  job aids, 148
  maintainability design for, 149
  nature of systems and, 137
  organizing and testing of, 152

206

power of systems approach for, 138
subsystem design specifications for, 142
subsystem outlining in, 141
system goals and measurements in, 138
training program design, 149
computer system parameter control as complex systems, 140
computer-based testing, 131
Consumer Product Safety Commission, product safety and, 171
contaminants, 45, 90, 183-184
contraction, muscular, 11
contrast, eyes, 7
controls, 119, 122
color coding for, 70
design principles for, 59-78, 59
discrete-action, 61
feedback from, 25
grouping of, 70
knob-type, 62
location of, 87
mode-of-operation coding for, 66-71
motion stereotypes for, 70
multidirectional tracking, 63
selection criteria for, 66-71
shape coding for, 70
size coding for, 70
unidimensional tracking, 61
conversion factors, metric measurements, 185
coordination, 4, 5, 12
cornea, 5
correlations, 181
counters, 64
CRT scope displays, 64
cutaneous receptors, 8

D

dark and light adaptation, eyes and, 6
decision games, 22
decision making, 21, 22
design
complex machines, 56
complex systems, 56
controls, 59
displays, 59
evaluation of, 155-167
handtools, 54
human factors in, 51-92
packaging, 54
simplex machines, 55
specifications for, 121
design evaluation, 155-167
drawing review, 157
environmental measurements and, 163-164
human factors affecting, 157-167
mockups for, 158-160
product improvement and, 163
prototyping and, 162
simulation assessments, 161
timeline analytical assessment techniques in, 161

deviation
average, 178
standard, 179
dial indicators, 64
digestive system, 13-14
discrete indicators, 64
discrete-action controls, 61
displays, 20, 122
auditory, 77
color impact on mood, emotion, appeal, 75
color in, 74
design principles for, 20, 59-78
labels, 76
location of, 87
mechanical signals, 64
preferred characteristics of, 65
reading distance in feet, 72
selection criteria for, 71-78
shared, 72
signs, 76
visual, 71
drawing review, design evaluation and, 157

E

ears, 5
effector subsystem, 9-13
limits to, 32
elderly, design limitations for, 48-49
emotional responses, 31
environmental stresses, 31, 38, 42-47, 89-91
complex systems and, 148
design evaluation and, 163-164
new products and, 120, 123
epilepsy, 44
equilibrium, 12
eyes, 5
afterimages, 6
binocular vision, 7
color blindness, 7
color sensitivity, 7
contrast, 7
dark and light adaptation, 6
limits to, 43-44
pattern vision, 7
retina of, 6

F

fatigue, 11, 42
feedback
improvements and product safety, 174
information, 27
manual controls, 25
flashes, 44
flicker, 43, 44
foolproofing, 174
functions
allocation of, 144
analysis of, 144

G

games, decision making, 22
gases (see also contaminants, pollutants), 45, 90
Gaussian distribution, 177
glare, 44
grip surfaces, 55

H

handicapped, design limitations for, 48-49
handtools (see tools)
hazardous chemicals, 91
heads-up display, 87, 88
Henry Dreyfuss Associates, 32
human characteristics, 15-50
  capabilities, 17-28
  decision making, 21, 22
  information processing, 21
  learning capacity, 27-28
  man and machine interaction, 19-26
  man vs. machine, 17-19
  management control, 18
  manual control, 24
  memory, 18
  multichannel perception, 18
  reading signals, 20
  reliability, 18-19
  versatility, 18
human factors design data, 51-92
  application of, 53-57
  basic principles of, 86
  biomechanical principles in, 88
  complex machines, 56
  complex systems, 56
  control and display design principles, 59-78
  environment, 89
  geometry of tools and packaging, 55
  grip surfaces, 55
  handtool design, 54
  layout design principles, 79-92
  packaging design, 54
  posture, 89
  simplex machines, 55
human form and function, 1-14
  autonomic nervous system, 13
  circulatory and digestive system, 13-14
  ears, 5
  eyes, 5
  muscles and muscular movement, 9-13
  nervous system and brain, 8
  senses, 3-8
human limitations, 29-50
  aged and impaired, 48-49
  biomechanical , 38-41
  children and adolescents, 47-48
  environment-related, 42-47
  gases and contaminants, 45
  information-processing, 30-31
  motion and acceleration, 46
  muscles and muscular movement, 32
  noise, 42
  sensory limitations, 29-30
  synergistic muscle action, 38-41
  temperature and humidity, 46
human operator (HO), 3
  machine interaction with, 19-26
human user (HU), 3
human-error, product safety and, 173
humidity, 46, 90

I

improvement feedback, 174
impulse noise, 42
information encoding and amplification, 31
information feedback, 27
information processing, 8, 21
  limits to, 30-31
infrasound, 42
inspection, maintenance and, 151
instruction (see training)
insurance
  human factors program for, 174
  product safety, 172
interaction of man and machine, 19-26
interfaces, complex systems and, 147

J

job aids, complex systems, 148

K

kinesthesis, 12
knob-type controls, 62

L

labels, 76, 166
  product safety and, 174
layout design principles, 79-92
  handtool design and, 80-83
  new products, 120, 123
  workstation design, 83-89
lead, 184
learning capacity, 27-28
left-handedness, 13
legal protection, product safety and, 174
lifting and handling, 38, 166
  force exerted in, 40
lighting conditions, 43, 44, 90
limitations (see human limitations)
litigation, product safety and, 172
loads (see lifting and handling)
long-term memory, 18

M

machines
  complex, 53
  simplex, 53
maintenance
  accessibility, 151
  complex systems, 149

Help Us Help You

So that we can better provide you with the practical information you need, please take a moment to complete and return this card.

1. **I am interested in books on the following subjects:**

☐ architecture & design
☐ automotive
☐ aviation
☐ business & finance
☐ computer, mini & mainframe
☐ computer, micros
☐ other_____

☐ electronics
☐ engineering
☐ hobbies & crafts
☐ how-to, do-it-yourself
☐ military history
☐ nautical

2. **I own/use a computer:**

☐ Apple/Macintosh_____
☐ Commodore_____
☐ IBM_____
☐ Other_____

3. **This card came from TAB book (no. or title):**

_____

4. **I purchase books from/by:**

☐ general bookstores
☐ technical bookstores
☐ college bookstores
☐ mail

☐ telephone
☐ electronic mail
☐ hobby stores
☐ art materials stores

Comments _____
_____
_____

Name _____

Address _____

City _____

State/Zip _____

TAB BOOKS Inc.